RECENT TOPICS IN NONLINEAR PDE II

This Page Intentionally Left Blank

NORTH-HOLLAND
MATHEMATICS STUDIES 128

Lecture Notes in Numerical and Applied Analysis Vol. 8

General Editors:

H. Fujita (University of Tokyo) and M. Yamaguti (Kyoto University)

Recent Topics in Nonlinear PDE II

Edited by

KYÛYA MASUDA (Tohoku University)
MASAYASU MIMURA (Hiroshima University)

KINOKUNIYA COMPANY LTD.
TOKYO JAPAN

NORTH-HOLLAND
AMSTERDAM·NEW YORK·OXFORD

KINOKUNIYA COMPANY-TOKYO
NORTH-HOLLAND-AMSTERDAM·NWE YORK·OXFORD

ISBN: 0 444 87938 2

Publishers
KINOKUNIYA COMPANY LTD.
TOKYO JAPAN

* * *

ELSEVIER SCIENCE PUBLISHERS B.V.
P.O. Box 1991
1000 BZ Amsterdam
The Netherlands

Sole distributors for the U.S.A. and Canada
ELSEVIER SCIENCE PUBLISHING COMPANY. INC.
52 VANDERBII.T AVENUE
NEW YORK. N.Y. 10017

Distributed in Japan by KINOKUNIYA COMPANY LTD.
Distributed outside Japan by ELSEVIER SCIENCE PUBLISHERS B. V. (NORTH-HOLLAND)

Lecture Notes in Numerical and Applied Analysis Vol. 8

PRINTED IN JAPAN

PREFACE

This volume is an outgrowth of lectures delivered at the second meeting on the subject of nonlinear partial differential equations, held at Tohoku University, February 27–29, 1984: The first meeting was held at Hiroshima University, 1983. The topics presented at the conference range over various fields in mathematical physics.

We would like to take the opportunity to thank all the participants of the meeting, and the contributors to this proceedings. Special thanks should go to Professors T. Muramatsu and J. Kato who helped in many ways to make the conference a success. We are also grateful to the Grant-in-Aid for Scientific Research from the Ministry of Education, Science and Culture of Japan for the financial support.

K. MASUDA
M. MIMURA

This Page Intentionally Left Blank

CONTENTS

This Page Intentionally Left Blank

Lecture Notes in Num. Appl. Anal., 8, 1–14 (1985)
Recent Topics in Nonlinear PDE II, Sendai, 1985

Large-Time Behavior of Viscous Surface Waves

by

J. Thomas BEALE* and Takaaki NISHIDA*

Duke University
Department of Mathematics
Durham, NC 27706

Kyoto University
Department of Mathematics
Kyoto, 606 Japan

§ 1 Introduction

We are concerned with global in time solutions to a free surface problem of the viscous incompressible fluid, which is formulated as follows: The motion of the fluid is governed by the Navier-Stokes equation

$$(1.1)\qquad \left.\begin{array}{r} u_t + (u\cdot\nabla)u - \nu\Delta u + \nabla p = 0 \\ \nabla\cdot u = 0 \end{array}\right\} \quad \text{in } \Omega(t) ,$$

where $\Omega(t) = \{ x \in R^2 , -b < y < \eta(t,x) \}$ is the domain occupied by the fluid. The free surface $S_F : y = \eta(t,x)$ satisfies the kinematic boundary condition

$$(1.2)\qquad \eta_t + u_1\eta_{x_1} + u_2\eta_{x_2} - u_3 = 0 \qquad \text{on } S_F .$$

The stress tensor satisfies the free boundary condition :

$$(1.3)\qquad pn_i - \nu(u_{i,x_j} + u_{j,x_i})n_j = [g\eta - \beta\nabla\{(1+|\nabla\eta|^2)^{-1/2}\nabla\eta\}]n_i \qquad \text{on } S_F ,$$

---

* Both authors are supported in part by the Mathematics Research Center, The University of Wisconsin-Madison.

where n is the outward normal to S_F, g is the gravitation constant and β is the nondimensionalized coefficient of surface tension. On the bottom $S_B : y = -b$ we have the fixed boundary condition

$$u = 0 \qquad \text{on } S_B . \tag{1.4}$$

We consider the initial value problem of (1.1)-(1.4) with the data at $t = 0$

$$\begin{cases} \eta = \eta_0(x) , & x \in R^2 , \\ u = u_0(x,y) & \text{in } \Omega_0 , \end{cases} \tag{1.5}$$

where $\Omega_0 = \Omega(0)$.

The local existence theorems for (1.1)-(1.5) are proved for both cases with or without considering the surface tension ([1],[2]). A global in time existence problem for (1.1)-(1.5) neglecting the surface tension (β=0) has a difficulty which was pointed out in [1]. However if the surface tension is taken into account, the following global existence and regularity theorem is proved.

Theorem 1.1 ([2])

Let $3 < r < 7/2$. Suppose the compatibility condition on the initial data :

$$\begin{cases} \nabla \cdot u_0 = 0 & \text{in } \Omega_0 , \\ \{((u_0)_{i,x_j} + (u_0)_{j,x_i})n_j\}_{\tan} = 0 & \text{on } y = \eta_0(x) , \\ u_0 = 0 & \text{on } y = -b . \end{cases} \tag{1.6}$$

There exists $\delta_0 > 0$ such that if the initial data

satisfy

$$(1.7) \quad E_0 \equiv |\eta_0|_{H^r(R^2)} + |u_0|_{H^{r-1/2}(\Omega_0)} < \delta_0 ,$$

then there exists a unique global solution η, u, p of (1.1)-(1.5), which satisfies

$$(1.8) \quad \eta \in \tilde{K}^{r+1/2}(R^+ \times R^2) , \; u \in K^r(R^+ \times \Omega(t)) , \; \nabla p \in K^{r-2}(R^+ \times \Omega(t)) .$$

Further given any $T_1 > 0$ and any $k > 0$, there exists $\delta_1 > 0$ such that if

$$(1.9) \quad E_0 < \delta_1$$

then the solution becomes smooth for $t > T_1$, i.e.,

$$(1.10) \quad \eta \in \tilde{K}^{r+k+1/2}((T_1,\infty)\times R^2) , \; u \subset K^{r+k}((T_1,\infty)\times\Omega(t)) ,$$

$$\nabla p \in K^{r+k-2}((T_1,\infty)\times\Omega(t)) .$$

In particular the solution with $k \gtrsim 2$ is classical. Here $H^r(\cdot\cdot)$ is the usual Sobolev space with norm $|\;|_r$ on the domain $\cdot\cdot$. $K^r((T_1,T_2)\times\Omega(t))$ is composed of the restriction to the fluid domain $\Omega(t)$ of the functions belonging to

$$(1.11) \; K^r((T_1,T_2)\times R^3) = H^0((T_1,T_2);H^r(R^3)) \cap H^{r/2}((T_1,T_2),H^0(R^3)) .$$

$\eta \in \tilde{K}^r(R^+\times R^2)$ is defined as follows :

$\eta \in K^r((0,T)\times R^2)$ for any $T > 0$ and $\eta = \eta_1 + \eta_2$ such that $\eta_1 \in K^r(R^+\times R^2)$ and η_2 is the Fourier transform in space-time of L^1 function of bounded support.

See [2] for the detailes of the function spaces.

In this summary we give an asymptotic decay rate for the

solution of the above theorem.

Theorem 1.2

If $\eta_0 \in L^1(R^2)$, then there exists $\delta_2 > 0$ such that if

$$(1.12) \qquad E_1 \equiv E_0 + |\eta_0|_{L^1} < \delta_2 ,$$

then the solution has the decay rate :

$$|\partial^\alpha \eta(t)|_0 < CE_1(1+t)^{-(1+\alpha)/2} , \quad \alpha = 0, 1, 2,$$

$$|u(t)|_2 , \quad |\nabla p(t)|_0 < CE_1(1+t)^{-1} .$$

In § 2 we transform the free boundary problem (1.1)-(1.5) to that on the fixed domain and reduce the components of the stress tensor to zero. The linear decay estimate is discussed in § 3 and the nonlinear one in § 4 .

§ 2 Reduction of the Problem

We remind ourselves some main ideas for the reduction of the free surface problem in [2]. First we use the transformation of the free boundary problem (1.1)-(1.5) to that on the fixed (equilibrium) domain : $\Omega = \{x \in R^2 , -b < y < 0\}$. Given $\eta(t,x)$ we extend it for $y < 0$ as follows :

$$(2.1) \qquad \tilde{\eta}(t,x,y) = \mathcal{F}^{-1}(e^{|\xi|y} \hat{\eta}(t,\xi)) ,$$

where $\hat{\eta}(t,\xi)$ is the Fourier transform with respect to x and $\mathcal{F}^{-1}$ is the inverse. If $\eta(t,\cdot)$ belongs to $H^s(S_F)$, then $\tilde{\eta}(t,\cdot,\cdot)$ belongs to $H^{s+1/2}(\Omega)$, where and hereafter S_F denotes the upper surface $y = 0$ of Ω . For each $t > 0$ we define the transformation θ on Ω onto $\Omega(t) = \{x \in R^2, -b < y < \eta(t,x)\}$ by

$$(2.2) \qquad \theta(x_1, x_2, y; t) = (x_1,\ x_2,\ \tilde{\eta}+y(1+\tilde{\eta}/b)) \quad .$$

The vector u on $\Omega(t) = \theta(\Omega)$ is defined from the vector v on Ω by

$$(2.3) \qquad u_i = \theta_{i,x_j} v_j \,/\, J \equiv \alpha_{ij}\, v_j \quad ,$$

where J is the Jacobian determinant of $d\theta = (\theta_{i,x_j}) = 1+\tilde{\eta}/b + \tilde{\eta}_y(1+y/b)$. This map conserves the property of divergence free. $\nabla\cdot v = 0$ in Ω iff $\nabla\cdot u = 0$ in $\Omega(t)$.

Using the transformation (2.2)(2.3) and $u_{i,x_j} = \zeta_{lj}\, \partial_l(\alpha_{ik}\, v_k)$, where $\zeta = (d\theta)^{-1}$ and so on, we can rewrite the free surface problem (1.1)-(1.5) to that on the equilibrium domain Ω as follows :

$$(2.4) \qquad \eta_t - v_3 = 0 \qquad \text{on } S_F \ ,$$

$$\left.\begin{aligned} &(2.5) \qquad v_t - \nu\,\Delta v + \nabla q = F(\eta,\ v,\ \nabla q) \\ &(2.6) \qquad \nabla\cdot v = 0 \end{aligned}\right\} \quad \text{in } \Omega \ ,$$

$$(2.7) \qquad v = 0 \qquad \text{on } S_B \ ,$$

$$\left.\begin{aligned} &(2.8) \qquad v_{i,x_3} + v_{3,x_i} = F_i(\eta,\ v) \ , \quad i = 1,\ 2 \\ &(2.9) \qquad q - 2\nu\, v_{3,x_3} - (g-\beta\Delta)\eta = F_3(\eta, v) \end{aligned}\right\} \quad \text{on } S_F$$

Here we have gathered the linear terms on the left hand side and all the nonlinear terms on the right hand side of the equation.

Next we reduce the tangential component of the stress tensor

F_i , $i = 1, 2$ to zero : Given $F_i \in H^{r-3/2}(S_F)$, $i = 1, 2$, choose the vector $z \in H^{r+1}(\Omega)$ satisfying the condition

$$\begin{cases} z = 0 , \quad \partial_y z = 0 , \quad \partial_y^{\,2} z = (F_2, -F_1, 0) & \text{on } S_F , \\ z = 0 , \quad \partial_y z = 0 & \text{on } S_B . \end{cases}$$

Then $w = \nabla\times z$ satisfies

$$w_3 = 0 , \quad w_{i,x_3} + w_{3,x_i} = F_i , \; i = 1, 2 \quad \text{on } S_F ,$$

$$\nabla\cdot w = 0 \quad \text{in } \Omega ,$$

$$w_3 = 0 \quad \text{on } S_F .$$

Therefore η, $v' = v - w$, q satisfy the system (2.4)-(2.9) with the replacements F by $F_4 = F - w_t + \nu\Delta w$ and F_i , $i = 1, 2$ by 0 . The prime of v' is omitted hereafter.

Last we rewrite the system (2.4)-(2.9) with $F = F_4$, $F_i = 0$, $i = 1, 2$, for η, v, q in the operator form. Let P be the projection on the subspace of solenoidal vectors orthogonal to the subspace $\mathcal{G}^0 = \{ \nabla\phi : \phi \in H^1(\Omega) , \phi = 0 \text{ on } S_F \}$ of $H^0(\Omega)$, i.e.,

$$(2.10) \qquad H^0 = PH^0 \oplus \mathcal{G}^0 .$$

Applying P to (2.5) we have

$$(2.11) \qquad v_t - \nu P\Delta v + P\nabla q = PF_4 .$$

Here $P\nabla q$ can be decomposed to three parts as follows :

$$P\nabla q = \nabla\pi^{(1)} + \nabla\pi^{(2)} + \nabla\pi^{(3)} ,$$

where $\pi^{(i)}$, $i = 1, 2, 3$, are defined by

$$\pi^{(1)} = 2\nu v_{3,x_3} \quad , \quad \pi^{(2)} = g\eta - \beta\Delta\eta \quad , \quad \pi^{(3)} = F_3 \quad \text{on} \quad S_F \quad ,$$

(2.12) $$\Delta \pi^{(i)} = 0 \quad \text{in} \quad \Omega \quad ,$$

$$\partial_y \pi^{(i)} = 0 \quad \text{on} \quad S_B \quad .$$

We denote

(2.13) $$\begin{cases} A\,v = -P\Delta v + \nabla\pi^{(1)} \quad , \\ R\,v = v_3|_{S_F} \quad , \\ R^*((g-\beta\Delta)\eta) = \nabla\pi^{(2)} \quad . \end{cases}$$

Using these notations the system (2.4),(2.11) has the form

(2.14) $$\eta_t = R\,v \quad ,$$

(2.15) $$v_t + A\,v + R^*((g-\beta\Delta)\eta) = f \quad ,$$

where $f(\eta, v, \nabla q) = PF_4 - \nabla\pi^{(3)}$.

(2.6)(2.7)(2.8) with $F_i = 0$ give the domain condition of A on v .

§ 3 Rates of Decay for Linear Problem

We investigate the decay rate of the solution of the linearized equation.

(3.1) $$\eta_t = Ru \quad ,$$

(3.2) $$u_t + Au + R^*((g-\beta\Delta)\eta) = 0 \quad ,$$

(3.3) $\eta(0) = \eta_0$, $u(0) = u_0$ at $t = 0$.

These are supplied with the conditions :

(3.4) $\nabla \cdot u = 0$ in Ω ,

(3.5) $u_{i,x_3} + u_{3,x_i} = 0$, $i = 1, 2$ on S_F ,

(3.6) $u = 0$ on S_B .

<u>Theorem 3.1</u> Let $E_2 = |\eta_0|_{L^1} + |\eta_0|_{5/2} + |u_0|_0$.

Then the solution of (3.1)-(3.6) has the decay rate :

$$|\partial^\alpha \eta(t)|_0 \leq C_0 E_2 \, t^{-(1+\alpha)/2} , \quad 0 \leq \alpha \leq 5/2 .$$
(3.7)
$$|\, u(t) \,|_2 \leq C_0 E_2 \, t^{-1} .$$

The theorem is proved by several steps.

Let $\mathcal{H} = \{ v = (\eta, u) : \eta \in H^1(S_F) , u \in PH^0(\Omega) \}$, where $(\rho, \eta)_1 = g(\rho, \eta)_0 + \beta(\nabla\rho, \nabla\eta)_0$ is the inner product of $H^1(S_F)$ and set $W = \{ v : \eta \in H^{5/2}(S_F) , u \in PH^2(\Omega)$ and u satisfies (3.4)(3.5)(3.6) $\}$. Let us define the operator

$$(3.9) \qquad G\,v = \begin{pmatrix} 0 & R \\ -R^*(g-\beta\Delta)\eta & -A \end{pmatrix} \begin{pmatrix} \eta \\ u \end{pmatrix} \qquad \text{from } D(G) \text{ in } \mathcal{H} ,$$

and consider its closed extension which will be denoted by G again.

<u>Lemma 3.2</u>

The operator G generates a contraction semigroup e^{tG} on $\mathcal{H}$, and $W \subset D(G)$.

Consider the resolvent equation :

$$(3.10) \qquad (\lambda - G)\begin{pmatrix} \eta \\ u \end{pmatrix} = \begin{pmatrix} h \\ f \end{pmatrix}$$

The resolvent of G can be extended to the left half plane as lemmas 3.3-3.5.

<u>Lemma 3.3</u> For any $\tau_0 > 0$ there exists $c_0 > 0$ such that if $\lambda \in \{ \lambda = \sigma + i\tau , -c_0|\tau| < \sigma < \tau_0 , |\tau| > \tau_0 \}$, then the solution of (3.10) has the estimate

$$(3.11) \qquad |u|_2 + |\lambda||u|_0 + |\lambda^{-1}Ru|_{5/2} + |\eta|_{5/2} + |\lambda||\eta|_{3/2} \le C(|f|_0 + |h|_{5/2}) .$$

We treat the resolvent near $\lambda = 0$ in two cases separately , i.e., (i) The supports of $\hat{h}(\xi)$, $\hat{f}(\xi,y)$ belong to $\{|\xi| \ge \xi_0\}$. (ii) The supports belong to $\{|\xi| \le \xi_0\}$. Here $\hat{}$ means the Fourier transform with respect to x .

<u>Lemma 3.4</u>

For any $\xi_0 > 0$ there exists $r_0 > 0$ such that if $\lambda \in \{ |\lambda| < r_0 \}$ and the supports of $\hat{h}(\xi)$, $\hat{f}(\xi,y)$ belong to $\{ |\xi| \ge \xi_0 \}$, then the resolvent equation (3.10) has the solution (η, u) satisfying

$$(3.12) \qquad |u|_2 , |\eta|_{5/2} \le C (|h|_{5/2} + |f|_0)$$

Let $\hat{G}(\xi)$ be the Fourier transform of G with respect to x .

<u>Lemma 3.5</u>

There exist $\xi_1 > 0$ and r_1, r_2 ($\nu(\pi/2b)^2 > r_2 > r_1 > 0$) such that if $r_1 < |\lambda| < r_2$ and $|\xi| < \xi_1$, then $(\lambda - \hat{G}(\xi))^{-1}$ exists except for a one-dimensional eigenspace which is analytic with respect to ξ . The eigenvalue and eigenvector have the following expansions.

$$(3.13)\quad \begin{cases} \lambda = -(gb^3/3\nu)|\xi|^2 + O(|\xi|^4) , \\ \eta = 1 + O(|\xi|^2) , \\ u_j = i(g\xi_j/2\nu)(y^2-b^2) + O(|\xi|^3) , \quad j = 1, 2 , \\ u_3 = (g|\xi|^2/2\nu)(y^3/3-b^2y-2b^3/3) + O(|\xi|^4) . \end{cases}$$

By using lemmas 3.2-3.5 the decay estimate (3.7) can be proved by the transformation of the integral path of the reprensentation

$$(3.14)\quad v(t) = e^{tG} v_0 = \lim_{\tau\to\infty} \frac{1}{2\pi i} \int_{\sigma-i\tau}^{\sigma+i\tau} e^{\lambda t} (\lambda - G)^{-1} v_0 \, d\lambda , \quad \sigma > 0$$

to the left half plane.

§ 4 Nonlinear Decay Estimates

The free surface problem (1.1) - (1.5) was reduced to the following system in § 2.

$$(4.1)\quad \eta_t - R u = 0 ,$$

$$(4.2)\quad u_t + A u + R^*((g-\beta\Delta)\eta) = f ,$$

$$(4.3)\quad \eta(0) = \eta_0 , \quad u(0) = u_0 ,$$

where f is nonlinear terms depending on η, u, ∇p and their derivatives.

The initial value problem (4.1) - (4.3) has the unique solution (Theorem 1.1) which becomes smooth for $t \geq T_1 > 0$. Namely we know that if $E_0 < \delta_1$, then

$$(4.4)\quad \begin{cases} \eta(t) \in C(0,\infty;H^{5/2}), \text{ i.e., } \quad \eta(t) \in C(0,\infty;H^3) \quad , \\ u(t) \in C(0,\infty;H^2), \text{ and} \end{cases}$$

$$(4.5)\qquad \tilde{\eta} \;\in\; \tilde{K}^7((1,\infty)\times\Omega) \quad , \quad u(t) \;\in\; K^6((1,\infty)\times\Omega) \quad .$$

Let us define

$$(4.6)\quad \begin{cases} M(\eta,u;t) = \max\{ \max\limits_{\alpha=0,1,2} \sup\limits_{0\le s\le t} (1+s)^{(1+\alpha)/2}|D^\alpha\tilde{\eta}(s)|_0 \quad , \\ \qquad \sup\limits_{0\le s\le t} (1+s)^{3/2}|D^3\tilde{\eta}(s)|_0 \quad , \quad \sup\limits_{0\le s\le t} (1+s)|u(s)|_2 \} \quad . \\ E(\eta,u) = |\tilde{\eta}|_{\tilde{K}^7((1,\infty)\times\Omega)} + |u|_{K^6((1,\infty)\times\Omega)} \quad . \end{cases}$$

Since the linearized equation is solved by semigroup e^{tG} on $\mathcal{H}$ in § 3, this solution satisfies the variation of constants formula:

$$(4.7)\qquad v(t) \;=\; e^{tG}\, v_0 + \int_0^t e^{(t-s)G}\, f(s)ds \quad ,$$

where $v = \binom{\eta}{u}$ and $f = \binom{0}{f(\eta,u,\nabla p)}$.

The first term of (4.7) has already estimated in § 3 ,i.e.,

$$(4.8)\qquad \begin{pmatrix}\eta_0(t)\\ u_0(t)\end{pmatrix} = e^{tG}\begin{pmatrix}\eta_0\\ u_0\end{pmatrix} \qquad \text{has the decay rate :}$$

$$|\partial^\alpha\eta_0(t)|_0 \;\le\; C_0E_2\, t^{-(1+\alpha)/2} \quad , \quad 0 \le \alpha \le 5/2 \text{ , i.e.,}$$

$$(4.9)\quad \begin{cases} |D^\alpha\tilde{\eta}_0(t)|_0 \;\le\; C_0E_2\, t^{-(1+\alpha)/2} \quad , \quad \alpha = 0,\, 1,\, 2 \text{ , and} \\ |D^3\tilde{\eta}_0(t)|_0 \;\le\; C_0E_2\, t^{-3/2} \quad , \\ |\, u_0(t)\, |_2 \;\le\; C_0E_2\, t^{-1} \quad . \end{cases}$$

It is sufficient by (4.4) to prove the decay rate for $t \ge 2$.

Let us decompose the second term of (4.7) into three parts

$$(4.10)\quad \int_0^t \{e^{(t-s)G}\begin{pmatrix}0\\ f(s)\end{pmatrix}\}ds = \int_0^{t/2}\{\quad\}ds + \int_{t/2}^{t-1}\{\quad\}ds + \int_{t-1}^{t}\{\quad\}ds$$

$$\equiv \begin{pmatrix}\eta_1\\ u_1\end{pmatrix}(t) + \begin{pmatrix}\eta_2\\ u_2\end{pmatrix}(t) + \begin{pmatrix}\eta_3\\ u_3\end{pmatrix}(t) .$$

<u>Lemma 4.1</u> Let $E_3 = |\tilde{\eta}|_3 + |u|_2 + |Dp|_0$. We have the estimates :

$$(4.11)\quad \begin{cases} |F|_0 \le C(E_3)\{(|D\tilde{\eta}|_2 + |u|_2)(|D\tilde{\eta}|_2 + |u|_2 + |Dp|_0) \\ \qquad + |\tilde{\eta}|_{L^\infty}(|D^2u|_0 + |Dp|_0)\}. \\ |DF|_0 \le C(E_3)\{(|D\tilde{\eta}|_2 + |u|_2)(|D\tilde{\eta}|_3 + |u|_3 + |Dp|_1) \\ \qquad + |\tilde{\eta}|_{L^\infty}(|D^3u|_0 + |D^2p|_0)\}. \\ |D^2F|_0 \le C(E_3)\{(|D\tilde{\eta}|_2 + |u|_2)(|D\tilde{\eta}|_4 + |u|_3 + |Dp|_1) \\ \qquad + (|\tilde{\eta}|_{L^\infty} + |D\tilde{\eta}|_{L^\infty} + |D^2\tilde{\eta}|_1)(|D^4u|_0 + |D^3p|_0)\}. \end{cases}$$

Let $\tilde{F}_i$, $i = 1, 2, 3$ on Ω be the extension of F_i on S_F by using $\tilde{\eta}$ for η . We have

$$(4.12)\quad |\tilde{F}_i|_0 \le C(E_3)(|D\tilde{\eta}|_2 + |\tilde{\eta}|_{L^\infty})|u|_2$$

$D^\alpha\tilde{F}_i$, $\alpha = 1, 2, 3,$ have the same estimate as $D^{\alpha-1}F$ as above. Therefore f has the same estimate as (4.11).

<u>Lemma 4. 2</u> For $t \ge 2$, we have for $i = 1, 2$

$$(4.13)\quad \begin{aligned} M(\eta_i,u_i;t) &\le C_1 \sup_{0\le s\le t}(1+s)^{3/2}|f(s)|_0 \\ &\le C_1C(E_3)M(\eta,u;t)^2 . \end{aligned}$$

Also we have for $i = 3$

$$(4.14) \quad \begin{aligned} M(\eta_3, u_3; t) &\leq C_2 \sup_{t-1 \leq s \leq t} (1 + s)^{3/2} |f(s)|_2 \\ &\leq C_2 C(E_3) M(\eta, u; t)(M(\eta, u; t) + E(\eta, u)) \quad . \end{aligned}$$

It is proved by Theorem 3.1 and by Lemma 4.1 and Sovolev-Nirenberg's inequality [4]. In particular, we know that since by (4.5) $\tilde{\eta}(t)$ and $u(t)$ are bounded in H^6 and H^5 respectively for any $t \geq 1$, we have by (4.6) for any $t \geq 1$

$$|D^{3+\alpha}\tilde{\eta}(t)|_0 \leq C\, t^{(\alpha-3)/2} \quad , \quad \alpha = 1, 2 ,$$

$$|D^{2+\alpha}u(t)|_0 \leq C\, t^{(\alpha/3-1)} \quad , \quad \alpha = 1, 2 .$$

Proof of Theorem 1.2

It follows from (4.9) and Lemma 4.2 that

$$\begin{aligned} M(\eta, u; t) &\leq C_0 E_2 + C_1 C M(\eta, u; t)^2 \\ &\quad + C_2 C M(\eta, u; t)(M(\eta, u; t) + E(\eta, u)) \quad . \end{aligned}$$

Therefore there exists $\delta_2 > 0$ such that if $E_2 < \delta_2$, $E_0 < \delta_1$, then

$$M(\eta, u; t) \leq C\, E_2 \quad .$$

This proves Theorem 1.2.

Remark. If the fluid has an infinite depth, the eigenvalue of $G(\xi)$ has the following expansion:

$$(4.15) \quad \lambda(\xi) = \pm\sqrt{g|\xi|} - 2\nu|\xi|^2 - O(|\xi|^{11/4}) \quad ,$$

which is quite different from (3.13). The details will be

published elsewhere.

References

[1] J. T. Beale, "The initial value problem for the Navier-Stokes equations with a free surface", Comm. Pure Appl. Math. 34 (1980), 359 - 392.

[2] J. T. Beale, "Large-Time Regularity of Viscous surface Waves", Arch. Rat. Mech. Anal. 84(1984), 307-352.

[3] T. Kato, "Perturbation Theory for Linear Operator", Springer-Verlag, Berlin-Heidelberg-New York, (1976)

[4] L. Nirenberg, "On elliptic partial differential equations", Annali della Scuola Norm. Sup. Pisa, 13(1959) 115-162

Lecture Notes in Num. Appl. Anal., 8, 15–52 (1985)
Recent Topics in Nonlinear PDE II, Sendai, 1985

On Representation of Solutions of Hamilton-Jacobi Equations with Convex Hamiltonians

Hitoshi ISHII
Department of Mathematics
Chuo University
Tokyo 112
Japan

§0. Introduction

Recently Crandall and Lions [5] introduced the notion of viscosity solution for Hamilton-Jacobi equations to settle the uniqueness problem of generalized solutions of Hamilton-Jacobi equations - see also Crandall-Evans-Lions [4] and Ishii [8]. The existence of viscosity solutions of Hamilton-Jacobi equations was established also under the same hypotheses on the Hamiltonians as those for the uniqueness of viscosity solutions. See Crandall-Lions [5], Lions [11,12], Souganidis [15], Barles [1] and Ishii [9].

In [11] Lions made the observation that the dynamic programming principle implies that the value function of an optimal control problem is the viscosity solution of its Bellman equation. It was proved by Souganidis [14], Barron-Evans-Jensen [2] and Evans-Souganidis [6] that the idea extends to the case of differential games. These results can be regarded that the viscosity solutions of the Bellman equation

$$u + \max_{a\in A} \{-g(x,a)\cdot Du - f(x,a)\} = 0 \qquad \text{in } \mathbb{R}^N \tag{0.1}$$

and of the Isaacs equation

$$u + \max_{a\in A} \min_{b\in B} \{-g(x,a,b)\cdot Du - f(x,a,b)\} = 0 \qquad \text{in } \mathbb{R}^N \tag{0.2}$$

are represented as the value functions, respectively, of an optimal control problem and of a differential game. In (0.1) and (0.2) Du denotes the gradient of u, $(\partial u/\partial x_1,\cdots,\partial u/\partial x_N)$. The representation results have weakness in the generality compared with the existence and uniqueness theorem for the viscosity solution of the stationary problem for the Hamilton-Jacobi equation

$$(0.3) \qquad u + H(x,Du) = 0 \quad \text{in } \mathbb{R}^N;$$

the representation theorems require much more on the Hamiltonian H.

We pose here a question: Is any viscosity solution the value function of an appropriate optimal control or differential game problem? We give a partial and positive answer to this question. Thus our goal of this paper is to represent the Hamiltonian $H(x,p)$ as a "max" or "max-min" of linear functions of p, i. e. to rewrite (0.3) in the form of (0.1) or (0.2), and to prove the uniform continuity of the value function of the associated optimal control or differential game problem. Then the dynamic programming principle and the uniqueness of the viscosity solution of (0.3) imply the value function is identical to the viscosity solution of (0.3). To carry out this program, we assume here that $p \to H(x,p)$ is convex. New difficulties arise when proving the uniform continuity of value functions and rewriting (0.3) in the form of (0.1). The former difficulty is resolved by introducing an argument analogous to the proof of the uniform continuity of viscosity solutions of Hamilton-Jacobi equations in [9] into optimal control theory. In the argument the continuity of the value function is derived from the continuity properties of the Hamiltonians with the help of selection lemmas (see e. g. Lemma 3.1). Its simplified version appears when we get a bound of the value functions (see §2). Our main tool to resolve the second difficulty is a uniform continuous selection lemma (see Lemma 5.1).

The plan of this paper is as follows. In Section 1 we introduce our control problem. We prove the uniform continuity of the value function of the optimal control problem in Sections 2 and 3. In Section 4 we observe that the dynamic programming principle implies that the value function satisfies the associated Bellman equation. In Section 5 we prove a representation theorem for convex Hamiltonians. In Section 6 we present a representation theorem for viscosity solutions of Hamilton-Jacobi equations with convex Hamiltonians.

We use the following notation throughout. $B(x,r) = B^N(x,r)$ denotes the closed ball in $\mathbb{R}^N$ of center x and radius $r \geq 0$. For $x, y \in \mathbb{R}^N$ $x \cdot y$ denotes the Euclidean inner product in $\mathbb{R}^N$. We will write $r^+ = \max\{r, 0\}$ for $r \in \mathbb{R}$. $UC(\mathbb{R}^N)$ and $BUC(\mathbb{R}^N)$ denote the spaces of uniformly continuous functions and of bounded, uniformly continuous functions on $\mathbb{R}^N$, respectively. For any set F, $\mathcal{P}(F)$ denotes the set of all subsets of F.

This work was begun when the author was visiting the Istituto Matematico, Università di Roma and supported in part by the (Italian) CNR. The author wishes to thank Prof. I. Capuzzo Dolcetta for his friendly hospitality. He also wishes to thank Prof. L. C. Evans for his criticism regarding the method of proof of the uniform continuity of value functions and Prof. R. Iino for useful comments on selection lemmas.

§1. An optimal control problem of infinite horizon

Let A be a nonempty metric space, and let $f: \mathbb{R}^N \times A \to \mathbb{R}$ and $g: \mathbb{R}^N \times A \to \mathbb{R}^N$ be given. We assume in the following that

(A1) f and g are continuous on $\mathbb{R}^N \times A$.

A (Lebesgue) measurable function $\alpha: [0,\infty) \to A$ is called a control and the set of controls is denoted by $\mathcal{A}$. We consider the ODE

$$(1.1) \quad \begin{cases} \dot{x}(t) = g(x(t),\alpha(t)) & \text{for a. a. } t \geq 0, \\ x(0) = x, \end{cases}$$

where $\alpha \in \mathcal{A}$ and $x \in \mathbb{R}^N$. By a solution of (1.1) we mean an absolutely continuous function $x(\cdot)$ which satisfies (1.1). We define

$$\mathcal{A}_g(x) = \{\alpha \in \mathcal{A} \mid (1.1) \text{ has a global solution}\}$$

for $x \in \mathbb{R}^N$. A control $\alpha \in \mathcal{A}_g(x)$ is called admissible at x if $t \to e^{-t} f(x(t),\alpha(t))^+$ is integrable on $[0,\infty)$ for a global solution $x(\cdot)$ of (1.1). $\mathcal{A}_{ad}(x)$ denotes the set of admissible controls at x. $X(x,\alpha)$ denotes the set of all solutions of (1.1) for $x \in \mathbb{R}^N$ and $\alpha \in \mathcal{A}$. For $x \in \mathbb{R}^N$ and $\alpha \in \mathcal{A}_{ad}(x)$, $x(\cdot) \in X(x,\alpha)$ is called admissible if $x(\cdot)$ is a global solution of (1.1) and satisfies the above integrability condition. The set of admissible $x(\cdot) \in X(x,\alpha)$ is denoted by $X_{ad}(x,\alpha)$. $X_g(x,\alpha)$ denotes the set of global solutions of (1.1).

An annoying point of our setting is that (A1) does not ensure in general the uniqueness of solutions of (1.1). This means that one can not control the system completely by selecting one of controls. This is, however, inevitable for us to get a general representation formula for viscosity solutions in view of the representation of Hamiltonians in §5 and also clarifies the effectiveness of our method of proof of the continuity of value functions.

Now we define the cost functional

$$(1.2) \qquad J(x,\alpha,x(\cdot)) = \int_0^\infty e^{-t} f(x(t),\alpha(t))dt$$

for $x \in \mathbb{R}^N$, $\alpha \in \mathcal{A}_{ad}(x)$, $x(\cdot) \in X_{ad}(x,\alpha)$ and the value function

(1.3) $$V(x) = \inf\{J(x,\alpha,x(\cdot)) \mid \alpha \in A_{ad}(x),\ x(\cdot) \in X_{ad}(x,\alpha)\}$$

for $x \in \mathbb{R}^N$.

This is an <u>infinite horizon</u> problem, and the goal is to find $V(x)$ and furthermore to find, if exist, a control $\alpha \in A_{ad}(x)$ and $x(\cdot) \in X_{ad}(x,\alpha)$ for which the infimum in (1.3) is achieved. Such a control is called an <u>optimal control</u>. Our contribution here is to demonstrate that $V \in UC(\mathbb{R}^N)$ and V solves (0.1) in the viscosity sense under assumptions (A1), (A2) and (A3) (see §2 for (A2) and (A3)).

§2. A bound for the value function

We will study the value function defined by (1.3) in this and the next two sections. Throughout these sections we assume (A1) and that the function $H: \mathbb{R}^N \times \mathbb{R}^N \to \mathbb{R}$ defined by

(2.1) $$H(x,p) = \sup_{a \in A}\{-g(x,a)\cdot p - f(x,a)\}$$

satisfies (A2) and (A3) listed below.

(A2) For each $R \geq 0$ there is a continuous function $\sigma_R: [0,2R] \to [0,\infty)$ satisfying $\sigma_R(0) = 0$ such that

$$|H(x,p) - H(x,q)| \leq \sigma_R(|p - q|)$$

for $x \in \mathbb{R}^N$ and $p, q \in B(0,R)$.

(A3) There is a continuous function $\omega: [0,\infty) \to [0,\infty)$ satisfying $\omega(0) = 0$ such that

$$|H(x,p) - H(y,p)| \leq \omega(|x-y|(1 + |p|))$$

for $x, y, p \in \mathbb{R}^N$.

We may assume that $(r,R) \to \sigma_R(r)$ is nondecreasing in each variable on $[0,2R] \times [0,\infty)$ and that

$$(2.2) \qquad \omega(r) \leq C_1(1 + r)$$

for all $r \geq 0$ and some constant $C_1 > 0$. We will write $\sigma(r) = \sigma_r(r)$ for $r \geq 0$. Note that $p \to H(x,p)$ is convex for $x \in \mathbb{R}^N$.

Remark 2.1. Conditions on H like (A2) and (A3) were employed by Crandall-Lions [5] when they formulated uniqueness results for viscosity solutions of (0.3) in the class $BUC(\mathbb{R}^N)$. It was later observed that conditions (A2) and (A3) are enough to ensure the existence and uniqueness of the viscosity solution of (0.3) in the class $UC(\mathbb{R}^N)$. See Ishii [8, 9], Lions [12], Souganidis [15] and Barles [1].

Theorem 2.1. Under assumptions (A1) - (A3), one has

$$(2.3) \qquad |V(x) + H(x,0)| \leq C_1 + \sigma(C_1) \qquad \text{for } x \in \mathbb{R}^N.$$

Corollary 2.1. Under assumptions (A1) - (A3),

$$(2.4) \qquad |V(x) - V(y)| \leq C_1|x - y| + 3C_1 + 2\sigma(C_1)$$

holds for $x, y \in \mathbb{R}^N$.

Proof. By (A3) and (2.2)

$$|H(x,0) - H(y,0)| \leq C_1|x - y| + C_1$$

for $x, y \in \mathbb{R}^N$. This and (2.3) together yield (2.4). Q.E.D.

To prove Theorem 2.1, we need the following lemma.

Lemma 2.1. For any $\varepsilon > 0$, $R > 0$ and $x \in \mathbb{R}^N$, there are $\alpha \in A_{ad}(x)$ and

$x(\cdot) \in X_{ad}(x,\alpha)$ such that

$$H(x(t),0) + f(x(t),\alpha(t)) < \varepsilon, \tag{2.5}$$

and

$$|x(t) - x| \leq \frac{t}{R}(\varepsilon + \sigma(R)) \tag{2.6}$$

for all $t \geq 0$.

Proof. Set

$$h(x,a) = H(x,0) + f(x,a) \qquad \text{for } x \in \mathbb{R}^N \text{ and } a \in A.$$

Since $\inf_{a \in A} h(x,a) = 0$ for $x \in \mathbb{R}^N$ by (2.1), one can select an $a_x \in A$ for each $x \in \mathbb{R}^N$ so that $h(x,a_x) < \varepsilon/2$. By the continuity of h there is an $r_x > 0$ for each $x \in \mathbb{R}^N$ such that

$$h(y,a_x) < \varepsilon \qquad \text{for } y \in B(x,r_x). \tag{2.7}$$

Setting

$$t_x = \frac{r_x}{2}\frac{R}{\varepsilon + \sigma(R)}$$

and letting $x(t;y)$ be any solution of (1.1) with $\alpha(t) \equiv a_x$ and the initial data y, we have

$$x(t;y) \in B(x,r_x) \tag{2.8}$$

for $y \in B(x,r_x/2)$ and $0 \leq t \leq t_x$. Indeed, supposing the contrary, we find $y_0 \in B(x,r_x/2)$ and $0 < t_0 < t_x$ such that

$$x(t_0;y_0) \in \partial B(x,r_x) \tag{2.9}$$

and

$$(2.10)\qquad x(t;y_0) \in B(x,r_x) \qquad \text{for all} \quad 0 \leq t \leq t_0.$$

It follows from (2.1) that $-g(y,a)\cdot p \leq H(y,p) + f(y,a)$ for $y,\ p \in \mathbb{R}^N$ and $a \in A$, and hence, by (A2),

$$R|g(y,a)| \leq H(y,-R\frac{g(y,a)}{|g(y,a)|}) + f(y,a) \leq H(y,0) + \sigma(R) + f(y,a)$$

if $g(y,a) \neq 0$. Thus by (2.7) and (2.10) we find that

$$R|\dot{x}(t;y_0)| = R|g(x(t;y_0),\alpha(t))| \leq \varepsilon + \sigma(R) \qquad \text{for a. a.} \quad 0 \leq t \leq t_0.$$

Hence

$$\begin{aligned} |x(t_0;y_0) - x| &\leq \int_0^{t_0} |\dot{x}(t;y_0)|dt + |x - y_0| \\ &\leq \frac{t_0}{R}(\varepsilon + \sigma(R)) + \frac{r_x}{2} < r_x; \end{aligned}$$

this contradicts (2.9). Thus (2.8) holds.

By compactness, for each $S > 0$, $B(0,S)$ can be covered with a finite subfamily of $\{B(x,r_x/2) \mid x \in B(0,S)\}$. Therefore it is possible to define mappings: $x \to a(x)$ from $\mathbb{R}^N$ to A and $r \to \tau(r)$ from $[0,\infty)$ to $(0,\infty)$ such that τ is nonincreasing and that

$$(2.11)\quad h(x(t),\alpha(t)) < \varepsilon \quad \text{if} \quad \alpha(t) \equiv a(x),\ x(\cdot) \in X(x,\alpha) \quad \text{and} \quad 0 \leq t \leq \tau(|x|).$$

Finally we fix $x \in \mathbb{R}^N$ and, using the mappings a and τ introduced above, define a sequence $\{(\alpha_j,\ \tau_j,\ x_j,\ y_j(\cdot))\}_{j\in\mathbb{N}}$ by the recursion formula

$$x_0 = x, \qquad \begin{cases} \alpha_j(t) \equiv a(x_{j-1}) \\ \tau_j = \tau(|x_{j-1}|) \\ y_j(\cdot) \in X_g(x_{j-1},\alpha_j) \\ x_j = y_j(\tau_j) \end{cases}$$

for $j \in \mathbb{N}$. Then we set

$$t_j = \tau_1 + \tau_2 + \cdots + \tau_j \qquad (j \in N),$$

$$\alpha(t) = \begin{cases} \alpha_1(t) & (0 \leq t < t_1) \\ \alpha_2(t - t_1) & (t_1 \leq t < t_2) \\ \vdots & \vdots \end{cases},$$

and

$$y(t) = \begin{cases} y_1(t) & (0 \leq t < t_1) \\ y_2(t - t_1) & (t_1 \leq t < t_2) \\ \vdots & \vdots \end{cases}.$$

We want to show that $t^* \equiv \lim_{j\to\infty} t_j = \infty$. To this end, we suppose $t^* < \infty$ and will obtain a contradiction. We have then $\lim_{j\to\infty} \tau_j = 0$. However, since $h(y(t),\alpha(t)) < \varepsilon$ for $0 \leq t < t^*$ by (2.11), it follows that $R|\dot{y}(t)| \leq \varepsilon + \sigma(R)$ for a. a. $0 \leq t < t^*$. Therefore

(2.12) $$|y(t) - x| \leq \frac{t}{R}(\varepsilon + \sigma(R)) \qquad \text{for } 0 \leq t \leq t^*.$$

This shows that $\tau_j \geq \tau(|x| + t^*(\varepsilon + \sigma(R))/R)$ for all $j \in \mathbb{N}$, which is a contradiction. Thus $\lim_{j\to\infty} t_j = \infty$, which proves that $\alpha \in A_g(x)$ and $y(\cdot) \in X_g(x,\alpha)$.

Since (2.11) implies (2.5) and (2.12) is exactly (2.6), it remains only to check that $t \to e^{-t} f(y(t),\alpha(t))^+$ is integrable on $[0,\infty)$. In view of (2.5), it is enough to verify that $t \to e^{-t} H(y(t),0)$ is integrable on $[0,\infty)$. By (A3), (2.2) and (2.12) we have

$$|e^{-t} H(y(t),0)| \leq e^{-t}\{|H(x,0)| + C_1(1 + |y(t) - x|)\}$$

$$\leq e^{-t}\{|H(x,0)| + C_1 + C_1(\varepsilon + \sigma(R))\tfrac{t}{R}\}$$

for $t \geqq 0$. The right side of this inequality is clearly integrable on $[0,\infty)$ as a function of t and so is its left side. Thus the proof is completed. Q.E.D.

Proof of Theorem 2.1. Let $\varepsilon > 0$ and $x \in \mathbb{R}^N$. By Lemma 2.1 there are $\alpha \in A_{ad}(x)$ and $x(\cdot) \in X_{ad}(x,\alpha)$ such that (2.5) and (2.6) with $R = C_1$ hold. We see then that

$$\begin{aligned} e^{-t} f(x(t),\alpha(t)) &< e^{-t}(\varepsilon - H(x(t),0)) \\ &\leqq e^{-t}(\varepsilon + C_1|x(t) - x| + C_1 - H(x,0)) \\ &\leqq e^{-t}\{\varepsilon + C_1 + (\varepsilon + \sigma(C_1))t - H(x,0)\} \end{aligned}$$

for $t \geqq 0$, and thus $J(x,\alpha,x(\cdot)) < 2\varepsilon + C_1 + \sigma(C_1) - H(x,0)$. Hence

$$V(x) + H(x,0) \leqq C_1 + \sigma(C_1) \qquad \text{for } x \in \mathbb{R}^N. \tag{2.13}$$

By (2.1) we have $H(x,p) \geqq -g(x,a)\cdot p - f(x,a)$ for x, $p \in \mathbb{R}^N$ and $a \in A$. Therefore, fixing $x \in \mathbb{R}^N$, $\alpha \in A_{ad}(x)$ and $x(\cdot) \in X_{ad}(x,\alpha)$, we find that

$$\begin{aligned} C_1|g(x(t),\alpha(t))| &\leqq H(x(t),0) + f(x(t),\alpha(t)) + \sigma(C_1) \\ &\leqq H(x,0) + C_1 + C_1|x(t) - x| + f(x(t),\alpha(t)) + \sigma(C_1) \end{aligned}$$

for $t \geqq 0$. Using this, we have

$$\begin{aligned} \frac{d}{dt}(e^{-t}|x(t) - x|) &\leqq (|\dot{x}(t)| - |x(t) - x|)e^{-t} \\ &\leqq (|g(x(t),\alpha(t))| - |x(t) - x|)e^{-t} \\ &\leqq \frac{1}{C_1}\{H(x,0) + C_1 + \sigma(C_1) + f(x(t),\alpha(t))\}e^{-t} \end{aligned}$$

for a. a. $t \geqq 0$. Integrating this over $[0,T]$, with $T > 0$, we get

$$\int_0^T e^{-t} f(x(t),\alpha(t))dt + H(x,0)(1 - e^{-T}) \geqq - C_1 - \sigma(C_1).$$

Taking into account that $t \to e^{-t}f(x(t),\alpha(t))^+$ is integrable on $[0,\infty)$, we find that $J(x,\alpha,x(\cdot)) + H(x,0) \geqq - C_1 - \sigma(C_1)$, from which we conclude that $V(x) + H(x,0) \geqq - C_1 - \sigma(C_1)$. This together with (2.13) proves (2.3) for $x \in \mathbb{R}^N$. Q.E.D.

Remark 2.2. Theorem 2.1 implies that if a measurable function $\alpha: [0,T] \to A$ and a solution $x(\cdot)$ of (1.1) defined on $[0,T]$, with $T > 0$, are given, then one can extend these functions to $[0,\infty)$ so that $\alpha \in A_{ad}(x)$ and $x(\cdot) \in X_{ad}(x,\alpha)$.

§3. Uniform continuity of the value function

The objective of this section is to prove the following

Theorem 3.1. Under assumptions (A1) - (A3), the value function V is uniformly continuous on $\mathbb{R}^N$. More precisely, there is a continuous function $\delta: [0,\infty) \to [0,\infty)$ depending only on ω in (A3) and σ_R in (A2) such that $\delta(0) = 0$, $\delta(r) > 0$ for $r > 0$ and

(3.1) $$|V(x) - V(y)| \leqq \varepsilon \quad \text{if } \varepsilon > 0 \text{ and } |x - y| \leqq \delta(\varepsilon).$$

We need a generalization of Lemma 2.1 for the proof of this theorem.

Lemma 3.1. Let $R > 0$, $\varepsilon > 0$ and $x \in \mathbb{R}^N$. Let $p: \mathbb{R}^N \times \mathbb{R}^N \times [0,\infty) \to B(0,R)$ be a function such that $t \to p(y,\eta,t)$ is measurable for $y,\eta \in \mathbb{R}^N$ and the family of functions: $(y,\eta) \to p(y,\eta,t)$, with $t \geqq 0$, is equicontinuous on $\mathbb{R}^N \times \mathbb{R}^N$. Then there are an absolutely continuous function $\xi: [0,\infty) \to \mathbb{R}^N$, sequences $\{a_i\}_{i\in\mathbb{N}} \subset A$ and $\{\lambda_i\}_{i\in\mathbb{N}} \subset L^\infty(0,\infty)$ such that the following conditions are satisfied:

(a) $\lambda_i(t) \geq 0 \quad (i \in \mathbb{N})$ and $\sum_{i\in\mathbb{N}} \lambda_i(t) = 1$ for $t \geq 0$.

(b) For each $T > 0$ there exists an $n_T \in \mathbb{N}$ such that $\lambda_i(t) = 0$ for $0 \leq t \leq T$ and $i > n_T$.

(c)
$$\sum_{i\in\mathbb{N}} \lambda_i(t)f(\xi(t),a_i) + H(\xi(t),p(\xi(t),\dot{\xi}(t),t)) + \dot{\xi}(t)\cdot p(\xi(t),\dot{\xi}(t),t) < \varepsilon,$$

$$\dot{\xi}(t) = \sum_{i\in\mathbb{N}} \lambda_i(t)g(\xi(t),a_i) \quad \text{and} \quad \xi(0) = x$$

hold for a. a. $t \geq 0$.

(d)
$$V(x) < \varepsilon + \int_0^\infty e^{-t} \sum_{i\in\mathbb{N}} \lambda_i(t)f(\xi(t),a_i)dt.$$

Remark 3.1. This lemma is closely related to the theory of relaxed controls. See e. g. Warga [16].

Remark 3.2. Condition (c) implies the integrability of

$$t \to e^{-t}\Big(\sum_{i\in\mathbb{N}} \lambda_i(t)f(\xi(t),a_i)\Big)^+$$

on $[0,\infty)$.

We will use the next lemma to prove Lemma 3.1.

Lemma 3.2. Let $T > 0$, $R > 0$, $\varepsilon > 0$ and $x \in \mathbb{R}^N$. Let p be as in Lemma 3.1. Then there exist an absolutely continuous function $\xi\colon [0,\infty) \to \mathbb{R}^N$, $\{a_i\}_{i\in\mathbb{N}} \subset A$, $\{\lambda_i\}_{i\in\mathbb{N}} \subset L^\infty(0,\infty)$, $\{\alpha_n\}_{n\in\mathbb{N}} \subset \mathcal{A}$ and a sequence $\{x_n(\cdot)\}_{n\in\mathbb{N}}$ of functions $x_n(\cdot) \in X(x,\alpha_n)$ fulfilling (a), (b) and (c) in Lemma 3.1 such that

(3.2)
$$\lim_{n\to\infty} \int_0^t e^{-s}f(x_n(s),\alpha_n(s))ds = \int_0^t e^{-s} \sum_{i\in\mathbb{N}} \lambda_i(s)f(\xi(s),a_i)ds$$

uniformly for $0 \leq t \leq T$ and

$$\lim_{n\to\infty} x_n(t) = \xi(t) \qquad \text{uniformly on } [0,T]. \tag{3.3}$$

The following lemma is needed to prove Lemma 3.2.

Lemma 3.3. Let $x \in \mathbb{R}^N$, $\{x_n\}_{n\in\mathbb{N}} \subset \mathbb{R}^N$, $\{a_i\}_{i=1}^m \subset A$, $T > 0$ and $r > 0$. Assume that $x_n \to x$ as $n \to \infty$ and that

$$T \sup\{|g(y,a_i)| \mid i = 1,\cdots,m,\ y \in B(x_n,r)\} \leq r \qquad \text{for } n \in \mathbb{N}. \tag{3.4}$$

Let $\{\lambda_i\}_{i=1}^m \subset L^\infty(0,\infty)$ satisfy

$$\lambda_i(t) \geq 0 \quad (i = 1,\cdots,m) \quad \text{and} \quad \sum_{i=1}^m \lambda_i(t) = 1 \quad \text{for } 0 \leq t \leq T.$$

Then there are an increasing sequence $\{n(k)\}_{k\in\mathbb{N}}$, $\{\alpha_k\}_{k\in\mathbb{N}} \subset A$, a sequence $\{x_k(\cdot)\}_{k\in\mathbb{N}}$ of functions $x_k(\cdot) \in X(x_{n(k)}, \alpha_k)$ defined on $[0,T]$ and an absolutely continuous function $\xi\colon [0,T] \to \mathbb{R}^N$ such that

$$\lim_{k\to\infty} \int_0^t e^{-s} f(x_k(s),\alpha_k(s))ds = \int_0^t e^{-s} \sum_{i=1}^m \lambda_i(s) f(\xi(s),a_i)ds \tag{3.5}$$

uniformly for $0 \leq t \leq T$,

$$\begin{cases} \dot{\xi}(t) = \sum_{i=1}^m \lambda_i(t) g(\xi(t),a_i) & \text{for a. a. } 0 \leq t \leq T, \\ \xi(0) = x, \end{cases} \tag{3.6}$$

and

$$\lim_{k\to\infty} x_k(t) = \xi(t) \qquad \text{uniformly on } [0,T]. \tag{3.7}$$

Remark 3.3. As the proof below shows, we can take $\{\alpha_k\}_{k\in\mathbb{N}}$ in this lemma so that $\alpha_k(t) \in \{a_1,\cdots,a_m\}$ for $t \geq 0$ and $k \in \mathbb{N}$.

Proof. For notational simplicity we assume $T = 1$. For any $n \in \mathbb{N}$

and $1 \leqq i \leqq m$, define $\lambda_i^{(n)} \in L^\infty(0,1)$ by

$$\lambda_i^{(n)}(t) = \frac{1}{n}\int_{(k-1)/n}^{k/n} \lambda_i(s)ds \quad \text{if} \quad \frac{k-1}{n} \leqq t < \frac{k}{n} \quad \text{for some} \quad k = 1,\cdots,n.$$

Then

$$\lambda_i^{(n)} \to \lambda_i \quad \text{in} \quad L^1(0,1) \tag{3.8}$$

for $i = 1,\cdots,m$ as $n \to \infty$ and

$$\lambda_i^{(n)}(t) \geqq 0 \quad (i = 1,\cdots,m), \qquad \sum_{i=1}^{m} \lambda_i^{(n)}(t) = 1 \tag{3.9}$$

for $0 \leqq t \leqq 1$ and $n \in \mathbb{N}$. Next we set

$$\tau_0^{(n)}(t) = 0, \qquad \tau_i^{(n)}(t) = \tau_{i-1}^{(n)}(t) + \lambda_i^{(n)}(t)/n \quad (i = 1,\cdots,m)$$

for $n \in \mathbb{N}$, $0 \leqq t < 1$ and define $\alpha_n \in A$ for $n \in \mathbb{N}$ by

$$\alpha_n(t) = \begin{cases} a_i & \text{if} \quad j/n + \tau_{i-1}^{(n)}(j/n) \leqq t < j/n + \tau_i^{(n)}(j/n) \\ & \text{for some} \quad j = 0,1,\cdots,n-1 \quad \text{and} \quad i = 1,\cdots,m, \\ a_1 & \text{if} \quad t \geqq 1. \end{cases} \tag{3.10}$$

For each $n \in \mathbb{N}$ we choose an $x_n(\cdot) \in X(x_n,\alpha_n)$ defined on $[0,T]$. The existence follows from (3.4) together with the standard local existence theorem. Moreover (3.4) guarantees that $\{x_n(\cdot)\}_{n\in\mathbb{N}}$ forms a uniformly bounded and equicontinuous family of functions on $[0,1]$. In view of the Ascoli-Arzela theorem we can extract a subsequence $\{x_{n(k)}(\cdot)\}_{k\in\mathbb{N}}$ of $\{x_n(\cdot)\}_{n\in\mathbb{N}}$ such that

$$\lim_{k\to\infty} x_{n(k)}(t) = \xi(t) \quad \text{uniformly on} \quad [0,1] \tag{3.11}$$

for some $\xi \in C([0,1], \mathbb{R}^N)$. Note here that

$$\int_0^t g(x_n(s),\alpha_n(s))ds = \sum_{j=0}^{[nt]-1}\sum_{i=1}^{m}\int_{j/n+\tau_{i-1}^{(n)}(j/n)}^{j/n+\tau_i^{(n)}(j/n)} g(x_n(s),a_i)ds$$

$$+\int_{[nt]/n}^{t} g(x_n(s),\alpha_n(s))ds$$

$$= \sum_{j=0}^{[nt]-1}\int_{j/n}^{(j+1)/n}\sum_{i=1}^{m}\lambda_i^{(n)}(s)g(x_n(s),a_i)ds + o(1)$$

$$= \int_0^t \sum_{i=1}^{m}\lambda_i^{(n)}(s)g(x_n(s),a_i)ds \quad + \quad o(1)$$

for all $n \in \mathbb{N}$ and $0 \leqq t \leqq 1$, where $o(1) \to 0$ uniformly for $0 \leqq t \leqq 1$ as $n \to \infty$ and $[nt]$ denotes the integral part of nt. Therefore, using (3.8) and (3.11), we have

$$\int_0^t g(x_n(s),\alpha_n(s))ds \quad \to \quad \int_0^t \sum_{i=1}^{m}\lambda_i(s)g(\xi(s),a_i)ds$$

as $n = n(k) \to \infty$, and hence

$$\xi(t) = x + \int_0^t \sum_{i=1}^{m}\lambda_i(s)g(\xi(s),a_i)ds \qquad \text{for all } 0 \leqq t \leqq 1.$$

This proves (3.6) and that ξ is absolutely continuous.

Similarly we see that

$$\int_0^t e^{-s}f(x_n(s),\alpha_n(s))ds \quad \to \quad \int_0^t e^{-s}\sum_{i=1}^{m}\lambda_i(s)f(\xi(s),a_i)ds$$

uniformly for $0 \leqq t \leqq 1$ as $n = n(k) \to \infty$. Thus we know that $\{\alpha_{n(k)}\}_{k\in\mathbb{N}} \subset A$, $\{x_{n(k)}(\cdot)\}_{k\in\mathbb{N}}$ and ξ have the required properties. Q.E.D.

Proof of Lemma 3.2. Step 1: We will choose $\{a_i\}_{i\in\mathbb{N}} \subset A$ and $\{\lambda_i(\cdot;x)\}_{i\in\mathbb{N}} \subset L^\infty(0,\infty)$ for all $x \in \mathbb{R}^N$. For simplicity we write

$$h(x,p,a) = H(x,p) + f(x,a) + g(x,a)\cdot p$$

for $x, p \in \mathbb{R}^N$ and $a \in A$.

Since $\inf_{a \in A} h(x,p,a) = 0$, fixing $\varepsilon > 0$, we can select $a_{x,p} \in A$ for each $x, p \in \mathbb{R}^N$ such that $h(x,p,a_{x,p}) < \varepsilon/2$. As $\{(y,q) \in \mathbb{R}^{2N} \mid h(y,q,a_{x,p}) < \varepsilon/2\}$ is open for $a, p \in \mathbb{R}^N$, we can choose a sequence $\{a_i\}_{i\in\mathbb{N}} \subset A$ such that

$$(3.12) \qquad B(0,S) \times B(0,R) \subset \bigcup_{i=1}^{m} \{(y,q) \mid h(y,q,a_i) < \frac{\varepsilon}{2}\}$$

for any $S > 0$ and some $m = m(S) \in \mathbb{N}$.

Now we fix $S > 0$, and let $m \in \mathbb{N}$ be such that (3.12) holds. Let $t \in [0,\infty)$ and $x \in B(0,S)$. We claim first that

(3.13) there exist $\lambda_1,\cdots,\lambda_m \in \mathbb{R}$ satisfying $\lambda_i \geq 0$ for all i and $\sum_{i=1}^{m} \lambda_i = 1$ such that

$$\sum_{i=1}^{m} \lambda_i h(x,p(x,\eta,t),a_i) \leq \frac{\varepsilon}{2},$$

where $\eta = \sum_{i=1}^{m} \lambda_i g(x,a_i)$.

To see this, we introduce the notation:

$$\Lambda = \{\lambda = (\lambda_1,\cdots,\lambda_m) \in \mathbb{R}^m \mid \lambda_i \geq 0 \text{ for all } i = 1,\cdots,m \text{ and } \sum_{i=1}^{m} \lambda_i = 1\},$$

$$\eta(\lambda) = \sum_{i=1}^{m} \lambda_i g(x,a_i) \qquad \text{for } \lambda \in \Lambda,$$

$$\psi(\lambda) = \{\gamma \in \Lambda \mid \sum_{i=1}^{m} \gamma_i h(x,p(x,\eta(\lambda),t),a_i) \leq \frac{\varepsilon}{2}\} \qquad \text{for } \lambda \in \Lambda.$$

Then, by (3.12), $\psi(\lambda) \neq \emptyset$ and is moreover a compact convex subset of $\mathbb{R}^m$ for $\lambda \in \Lambda$. It follows from the continuity of η, $p(x,\cdot,t)$ and h that $\psi: \Lambda \to \mathcal{P}(\Lambda)$ is upper semicontinuous (see Kakutani [10]). Now we apply Kakutani's fixed point theorem [10] to ψ and find $\lambda \in \Lambda$ such that $\lambda \in \psi(\lambda)$. That is, (3.13) holds.

Now, keeping $x \in B(0,S)$ fixed, we show that

(3.14) there exists a measurable $\lambda\colon [0,\infty) \to \Lambda$ such that

$$\sum_{i=1}^{m} \lambda_i(t)h(x,p(x,\eta(t),t),a_i) \leq \frac{\varepsilon}{2} \qquad \text{for } t \geq 0,$$

where $\eta(t) = \sum_{i=1}^{m} \lambda_i(t)g(x,a_i)$ and $\lambda(t) = (\lambda_1(t),\cdots,\lambda_m(t))$.

To this end, we define $\tilde{h}\colon \Lambda \times [0,\infty) \to \mathbb{R}$ and $\phi\colon [0,\infty) \to P(\mathbb{R}^m)$ by

$$\tilde{h}(\lambda,t) = \sum_{i=1}^{m} \lambda_i h(x,p(x,\sum_{j=1}^{m} \lambda_j g(x,a_j),t),a_i),$$

$$\phi(t) = \{\lambda \in \Lambda | \tilde{h}(\lambda,t) \leq \frac{\varepsilon}{2}\} .$$

Since $\lambda \to \tilde{h}(\lambda,t)$ is continuous and $t \to \tilde{h}(\lambda,t)$ is measurable, $\phi\colon [0,\infty) \to P(\mathbb{R}^m)$ is measurable in the sense of [3, §3.1]. $\phi(t)$ is closed and, by (3.13), nonempty for $t \geq 0$. Hence by a measurable selection lemma (see, e. g., [3, Theorem 3.1.1]) there exists a measurable function $\lambda\colon [0,\infty) \to \mathbb{R}^m$ such that $\lambda(t) \in \phi(t)$ for all $t \geq 0$. This inclusion is equivalent to the inequality in (3.14). Thus we conclude (3.14).

The observation (3.14), the continuity of h and $p(\cdot,\cdot,t)$ and the standard compactness argument together imply that, for each $x \in \mathbb{R}^N$, there exist $r(x) \in (0,1)$, $m(x) \in \mathbb{N}$ and measurable functions $\lambda_i(\cdot;x)\colon [0,\infty) \to [0,\infty)$ $(i = 1,\cdots,m(x))$ having the properties listed below:

(3.15) $$\sum_{i=1}^{m(x)} \lambda_i(t;x) = 1 \qquad \text{for } t \geq 0 \text{ and } x \in \mathbb{R}^N.$$

(3.16) $$\sum_{i=1}^{m(x)} \lambda_i(t;x)h(y,p(y,\eta(x,y,t),t),a_i) < \varepsilon$$

for $t \geq 0$, $x \in \mathbb{R}^N$ and $y \in B(x,r(x))$, where

$$\eta(x,y,t) = \sum_{i=1}^{m(x)} \lambda_i(t;x)g(y,a_i).$$

(3.17) $$\inf_{x \in B(0,S)} r(x) > 0 \qquad \text{for } S > 0.$$

$$\sup_{x \in B(0,S)} m(x) < \infty \quad \text{for } S > 0. \tag{3.18}$$

Choose a continuous function $\tau: \mathbb{R}^N \to (0,\infty)$ such that

$$\tau(x) \sup \{|g(y,a_i)|\mid 1 \leqq i \leqq m(x),\ y \in B(x,r(x))\} < r(x)$$

for $x \in \mathbb{R}^N$. Note that if $s \geqq 0$ and $\xi \in C([0,\infty), \mathbb{R}^N)$ satisfies

$$\xi(t) = x + \int_s^t \sum_{i=1}^{m(x)} \lambda_i(u;x) g(\xi(u),a_i)du \quad \text{for } s \leqq t \leqq s + \tau(x),$$

then $\xi(t) \in B(x,r(x))$ for $s \leqq t \leqq s + \tau(x)$.

Step 2: We will construct $\{\lambda_i\}_{i \in \mathbb{N}} \subset L^\infty(0,\infty)$ and an absolutely continuous function $\xi: [0,\infty) \to \mathbb{R}^N$ satisfying conditions (a), (b) and (c) in Lemma 3.1 for $x \in \mathbb{R}^N$. To do this, we use a step-by-step argument.

Let $x \in \mathbb{R}^N$, and define $\{\tau_j\}_{j \in \mathbb{N}} \subset (0,\infty)$, $\{\lambda_i\}_{i \in \mathbb{N}} \subset L^\infty(0,\infty)$, $\xi \in C([0,\infty), \mathbb{R}^N)$ and $\{\xi_j\}_{j \in \mathbb{N}} \subset \mathbb{R}^N$ as follows. Set $\tau_1 = \tau(x)$ and

$$\lambda_i(t) = \begin{cases} \lambda_i(t;x) & \text{for } 1 \leqq i \leqq m(x), \\ 0 & \text{for } i > m(x) \end{cases}$$

for $0 \leqq t < \tau_1$. Let $\xi \in C([0,\tau_1], \mathbb{R}^N)$ be a solution of $\xi(t) = x + \int_0^t \sum_{i=1}^\infty \lambda_i(s) g(\xi(s),a_i)ds$ for $0 \leqq t \leqq \tau_1$. (Its existence follows from Lemma 3.3 or the argument used to verify the existence of $x_n(\cdot)$ in the proof of Lemma 3.3.) Set $\xi_1 = \xi(\tau_1)$. It follows that $\xi(t) \in B(x,r(x))$ for $0 \leqq t \leqq \tau_1$. Hence, by (3.16),

$$\sum_{i=1}^\infty \lambda_i(t) h(\xi(t), p(\xi(t),\eta(t),t), a_i) < \varepsilon \tag{3.19}$$

for $0 \leqq t < \tau_1$, where $\eta(t) = \sum_{i=1}^\infty \lambda_i(t) g(\xi(t),a_i)$.

Next we set $\tau_2 = \tau_1 + \tau(\xi_1)$ and

$$\lambda_i(t) = \begin{cases} \lambda_i(t;\xi_1) & \text{if } 1 \le i \le m(\xi_1), \\ 0 & \text{if } i > m(\xi_1) \end{cases}$$

for $\tau_1 \le t < \tau_2$. Extend ξ to $[0,\tau_2]$ so that $\xi \in C([0,\tau_2], \mathbb{R}^N)$ and $\xi(t) = x + \int_0^t \sum_{i=1}^\infty \lambda_i(s)g(\xi(s),a_i)ds$ holds for $0 \le t \le \tau_2$. We set $\xi_2 = \xi(\tau_2)$. As above, we see that (3.19) holds for $0 \le t < \tau_2$. Repeat this procedure to obtain $\{\tau_j\}_{j\in\mathbb{N}} \subset (0,\infty)$, $\{\lambda_i\}_{i\in\mathbb{N}} \subset L^\infty(0,\tau^*)$ and $\xi \in C([0,\tau^*), \mathbb{R}^N)$, where $\tau^* = \lim_{j\to\infty} \tau_j$. Then (3.19) holds for $0 \le t < \tau^*$. Notice that if $\tau^* = \infty$, then $\{\lambda_i\}_{i\in\mathbb{N}}$ satisfies conditions (a) and (b) and (3.19) is equivalent to the first inequality in (c) since $\xi(t) = x + \int_0^t \sum_{i=1}^\infty \lambda_i(s)g(\xi(s),a_i)ds$ for all $t \ge 0$.

Thus, it is enough to show that $\tau^* = \infty$. To the contrary, we suppose $\tau^* < \infty$. This implies $\tau(\xi_j) \to 0$ as $j \to \infty$. Combine the inequality

$$H(\xi(t),q) \ge -\sum_{i=1}^{\infty} \lambda_i(t)\{g(\xi(t),a_i)\cdot q + f(\xi(t),a_i)\}$$

(which follows from the definition of H), with $q = p(\xi(t),\dot\xi(t),t) - \dot\xi(t)/|\dot\xi(t)|$, and the first inequality in (c) to get

$$|\dot\xi(t)| \le \varepsilon + \sigma(R+1) \qquad \text{for } 0 \le t < \tau^*.$$

Therefore, $|\xi(t) - x| \le (\varepsilon + \sigma(R+1))\tau^*$ for $0 \le t < \tau^*$, and hence $\inf_{j\in\mathbb{N}} \tau(\xi_j) = \inf_{j\in\mathbb{N}} \tau(\xi(\tau_j)) > 0$. This is a contradiction, which shows $\tau^* = \infty$.

Step 3: For any $T > 0$ we will choose $\xi \in C([0,\infty), \mathbb{R}^N)$ and $\{\lambda_i\}_{i\in\mathbb{N}} \subset L^\infty(0,\infty)$ satisfying (a), (b) and (c) so that ξ can be approximated by solutions of (1.1) on $[0,T]$.

Fix $x \in \mathbb{R}^N$. We define $\tau^{(j)} > 0$, $\{\lambda_i^{(j)}\}_{i\in\mathbb{N}} \subset L^\infty(0,\tau^{(j)})$, $\xi^{(j)} \in C([0,\tau^{(j)}], \mathbb{R}^N)$, $\xi_j \in \mathbb{R}^N$, $\{\alpha_n^{(j)}\}_{n\in\mathbb{N}} \subset A$ and a sequence $\{x_n^{(j)}(\cdot)\}_{n\in\mathbb{N}}$ of functions $x_n^{(j)}(\cdot) \in X(x,\alpha_n^{(j)})$ on $[0,\tau^{(j)}]$ for every $j \in \mathbf{N}$ in the following manner. Set $\tau^{(1)} = \tau(x)$ and

$$\lambda_i^{(1)}(t) = \begin{cases} \lambda_i(t;x) & \text{if } 1 \leq i \leq m(x), \\ 0 & \text{if } i > m(x) \end{cases}$$

for $0 \leq t < \tau^{(1)}$. By virtue of Lemma 3.3, there exist $\{\alpha_n^{(1)}\}_{n\in\mathbf{N}} \subset A$, a sequence $\{x_n^{(1)}(\cdot)\}_{n\in\mathbf{N}}$ of functions $x_n^{(1)}(\cdot) \in X(x,\alpha_n^{(1)})$ on $[0,\tau^{(1)}]$ and $\xi^{(1)} \in C([0,\tau^{(1)}]; \mathbf{R}^N)$ satisfying the following conditions $(3.20)_j$, $(3.21)_j$ and $(3.22)_j$ for $j = 1$.

$$(3.20)_j \quad \lim_{n\to\infty} \int_0^t e^{-s} f(x_n^{(j)}(s),\alpha_n^{(j)}(s))ds = \int_0^t e^{-s} \sum_{i=1}^{\infty} \lambda_i^{(j)}(s) f(\xi^{(j)}(s),a_i)ds$$

uniformly for $0 \leq t \leq \tau^{(j)}$.

$$(3.21)_j \quad \xi^{(j)}(t) = x + \int_0^t \sum_{i=1}^{\infty} \lambda_i^{(j)}(s) g(\xi^{(j)}(s),a_i)ds \quad \text{for } 0 \leq t \leq \tau^{(j)}.$$

$$(3.22)_j \quad \lim_{n\to\infty} x_n^{(j)}(t) = \xi^{(j)}(t) \quad \text{uniformly on } [0,\tau^{(j)}].$$

As in Step 2, we have

$$(3.23) \quad \sum_{i=1}^{\infty} \lambda_i^{(1)}(t) h(\xi^{(1)}(t),p(\xi^{(1)}(t),\eta(t),t),a_i) < \varepsilon$$

for $0 \leq t \leq \tau^{(1)}$, where $\eta(t) = \sum_{i=1}^{\infty} \lambda_i^{(1)}(t) g(\xi^{(1)}(t),a_i)$. We put $\xi_1 = \xi^{(1)}(\tau^{(1)})$.

Now we set $\tau^{(2)} = \tau^{(1)} + \tau(\xi_1)$ and

$$\lambda_i^{(2)}(t) = \begin{cases} \lambda_i^{(1)}(t) & \text{for } 0 \leq t < \tau^{(1)}, \\ \lambda_i(t;\xi_1) & \text{for } 1 \leq i \leq m(\xi_1) \text{ and } \tau^{(1)} \leq t < \tau^{(2)}, \\ 0 & \text{for } i > m(\xi_1) \text{ and } \tau^{(1)} \leq t < \tau^{(2)}. \end{cases}$$

Applying Lemma 3.3 with $\{x_n\}_{n\in\mathbf{N}} = \{x_n^{(1)}(\tau^{(1)})\}_{n\in\mathbf{N}}$ and the interval $[0,T]$ replaced by $[\tau^{(1)},\tau^{(2)}]$, we find that there exist $\{\alpha_n^{(2)}\}_{n\in\mathbf{N}} \subset A$, a sequence

$\{x_n^{(2)}(\cdot)\}_{n\in\mathbb{N}}$ of functions $x_n^{(2)}(\cdot) \in X(x,\alpha_n^{(2)})$ on $[0,\tau^{(2)}]$ and $\xi^{(2)}$ $\in C([0,\tau^{(2)}], \mathbb{R}^N)$ such that $(3.20)_2$, $(3.21)_2$, $(3.22)_2$ and $\xi^{(2)}(\tau^{(1)}) = \xi_1$ are satisfied. The inequality

$$(3.24)_j \qquad \sum_{i=1}^{\infty} \lambda_i^{(j)}(t)g(\xi^{(j)}(t),p(\xi^{(j)}(t),\eta^{(j)}(t),t),a_i) < \varepsilon \qquad \text{for } 0 \leq t < \tau^{(j)}$$

is valid for $j = 2$, where $\eta^{(j)}(t) = \sum_{i=1}^{\infty} \lambda_i^{(j)}(t)g(\xi^{(j)}(t),a_i)$. Lastly we put $\xi_2 = \xi^{(2)}(\tau^{(2)})$.

Repeating in this way, for any $j \in \mathbb{N}$, we can define $\tau^{(j)} > 0$, $\{\lambda_i^{(j)}\}_{i\in\mathbb{N}}$ $\subset L^\infty(0,\tau^{(j)})$, $\xi^{(j)} \in C([0,\tau^{(j)}], \mathbb{R}^N)$, $\{\alpha_n^{(j)}\}_{n\in\mathbb{N}} \subset A$ and a sequence $\{x_n^{(j)}(\cdot)\}_{n\in\mathbb{N}}$ of functions $x_n^{(j)}(\cdot) \in X(x,\alpha_n^{(j)})$ on $[0,\tau^{(j)}]$ so that $(3.20)_j$, $(3.21)_j$, $(3.22)_j$ and $(3.24)_j$ hold. *It is verified as in Step 2 that*

$$\lim_{j\to\infty} \tau^{(j)} = \infty.$$

What we saw in Step 2 is that we can extend $\lambda_i^{(j)}$, $\xi^{(j)}$ to $[0,\infty)$ so that conditions (a), (b) and (c) are satisfied for the extended $\{\lambda_i^{(j)}\}_{i\in\mathbb{N}}$ and $\xi^{(j)}$. Thus we conclude that, for any $T > 0$, there exist $\xi \in C([0,\infty), \mathbb{R}^N)$, $\{\lambda_i\}_{i\in\mathbb{N}} \subset L^\infty(0,\infty)$, $\{\alpha_n\}_{n\in\mathbb{N}} \subset A$ and a sequence $\{x_n(\cdot)\}_{n\in\mathbb{N}}$ of functions $x_n(\cdot)$ belonging to $X(x,\alpha_n)$ such that ξ and $\{\lambda_i\}$ satisfy (a), (b) and (c) and that

$$(3.25) \qquad \lim_{n\to\infty} \int_0^t e^{-s}f(x_n(s),\alpha_n(s))ds = \int_0^t e^{-s} \sum_{i=1}^{\infty} \lambda_i(s)f(\xi(s),a_i)ds$$

uniformly for $0 \leq t \leq T$ and

$$(3.26) \qquad \lim_{n\to\infty} x_n(t) = \xi(t) \qquad \text{uniformly on } [0,T].$$

The proof is now completed. Q.E.D.

Proof of Lemma 3.1. Let $\xi \in C([0,\infty), \mathbb{R}^N)$ and $\{\lambda_i\}_{i\in\mathbb{N}} \subset L^\infty(0,\infty)$ satisfy (a), (b) and (c). As in Step 2 of the proof of Lemma 3.2, we have

$$|\dot{\xi}(t)| \leq \varepsilon + \sigma(R+1) \qquad \text{for a. a.} \quad t \geq 0,$$

and so

$$(3.27) \qquad |\xi(t) - x| \leq (\varepsilon + \sigma(R+1))t \qquad \text{for} \quad t \geq 0.$$

These, together with (c) and that $H(\xi(t),0) \geq -\sum_{i=1}^{\infty} \lambda_i(t)f(\xi(t),a_i)$ for $t \geq 0$, yield

$$\left|\sum_{i=1}^{\infty} \lambda_i(t)f(\xi(t),a_i) + H(x,0)\right|$$
$$\leq \varepsilon + (\varepsilon + \sigma(R+1))R + \sigma(R) + C_1 + C_1(\varepsilon + \sigma(R+1))t$$

for a. a. $t \geq 0$. This shows that there is a $T > 0$ independent of the choice of ξ and $\{\lambda_i\}_{i\in\mathbb{N}}$ such that

$$(3.28) \qquad \left|\int_T^\infty e^{-t} \sum_{i=1}^{\infty} \lambda_i(t)f(\xi(t),a_i)dt\right| < \frac{\varepsilon}{4}.$$

Replacing $T > 0$ by a larger number if necessary, we assume in addition that

$$(3.29) \qquad e^{-T}\{|H(0,0)| + C_1(|x| + \varepsilon T + \sigma(R+1)T + 3) + \sigma(C_1)\} < \frac{\varepsilon}{4}.$$

By virtue of Lemma 3.2, we can select $\xi \in C([0,\infty), \mathbb{R}^N)$, $\{\lambda_i\}_{i\in\mathbb{N}} \subset L^\infty(0,\infty)$, $\alpha \in A$ and $x(\cdot) \in X(x,\alpha)$ defined on $[0,T]$ so that (a), (b) and (c) are satisfied and that

$$(3.30) \qquad \left|\int_0^t e^{-s}\{f(x(s),\alpha(s)) - \sum_{i=1}^{\infty} \lambda_i(s)f(\xi(s),a_i)\}ds\right| < \frac{\varepsilon}{4}$$

and

$$(3.31) \qquad |x(t) - \xi(t)| \leq 1$$

hold for $0 \leq t \leq T$. Choose $\beta \in A_{ad}(x(T))$ and $y(\cdot) \in X_{ad}(x(T),\beta)$ so that

$$J(x(T),\beta,y(\cdot)) < \frac{\varepsilon}{4} + V(x(T)).$$

Then, by Theorem 2.1, (3.27) and (3.31), we get

$$J(x(T),\beta,y(\cdot)) < \frac{\varepsilon}{4} + |H(0,0)| + C_1(|x| + \varepsilon T + \sigma(R+1)T + 3) + \sigma(C_1). \tag{3.32}$$

Now we define

$$\tilde{\alpha}(t) = \begin{cases} \alpha(t) & \text{for } 0 \leq t \leq T, \\ \beta(t - T) & \text{for } t > T \end{cases}$$

and

$$\tilde{x}(t) = \begin{cases} x(t) & \text{for } 0 \leq t \leq T, \\ y(t - T) & \text{for } t > T. \end{cases}$$

Note that $\tilde{x}(\cdot) \in X_{ad}(x,\tilde{\alpha})$ and that

$$J(x,\tilde{\alpha},\tilde{x}(\cdot)) = \int_0^T e^{-t} f(x(t),\alpha(t))dt + e^{-T} J(x(T),\beta,y(\cdot)).$$

Combining this with (3.30), (3.32), (3.28) and (3.29) yields

$$V(x) \leq J(x,\tilde{\alpha},\tilde{x}(\cdot)) < \varepsilon + \int_0^\infty e^{-t} \sum_{i=1}^\infty \lambda_i(t) f(\xi(t),a_i)dt.$$

Thus we have completed the proof. Q.E.D.

The dynamic programming principle is stated as follows.

<u>Proposition 3.1</u>. For any $t > 0$ and $x \in \mathbb{R}^N$ one has

$$V(x) = \inf \{\int_0^t e^{-s} f(x(s),\alpha(s))ds + e^{-t} V(x(t))\}, \tag{3.33}$$

where the infimum is taken over all $\alpha \in A_{ad}(x)$, $x(\cdot) \in X_{ad}(x,\alpha)$.

We refer to Lions [11] for a proof of Proposition 3.1.

Proposition 3.2. Let $T > 0$, $R > 0$, $\varepsilon > 0$, $x \in \mathbb{R}^N$, and let p be as in Lemma 3.1. Then there exist an absolutely continuous function $\xi: [0,\infty) \to \mathbb{R}^N$, $\{a_i\}_{i\in\mathbb{N}} \subset A$ and $\{\lambda_i\}_{i\in\mathbb{N}} \subset L^\infty(0,\infty)$ satisfying (a), (b) and (c) in Lemma 3.1 such that

$$(3.34)\quad V(x) \leq \int_0^t e^{-s} \sum_{i=1}^\infty \lambda_i(s) f(\xi(s),a_i)ds + e^{-t}V(\xi(t)) + e^{-t}(3C_1 + 2\sigma(C_1))$$

for $0 < t \leq T$.

Proof. By Lemma 3.2, we can choose an absolutely continuous function ξ: $[0,\infty) \to \mathbb{R}^N$, $\{a_i\}_{i\in\mathbb{N}} \subset A$, $\{\lambda_i\}_{i\in\mathbb{N}} \subset L^\infty(0,\infty)$, $\{\alpha_n\}_{n\in\mathbb{N}} \subset A_{ad}(x)$ and $\{x_n(\cdot)\}_{n\in\mathbb{N}}$, where $x_n(\cdot) \in X_{ad}(x,\alpha_n)$, satisfying (a), (b) and (c) such that

$$(3.35)\qquad \left|\int_0^t e^{-s}\{f(x_n(s),\alpha_n(s)) - \sum_{i=1}^\infty \lambda_i(s)f(\xi(s),a_i)\}ds\right| < \frac{1}{n}$$

and

$$(3.36)\qquad |x_n(t) - \xi(t)| < \frac{1}{n}$$

hold for $0 \leq t \leq T$ and $n \in \mathbb{N}$. By Proposition 3.1, we find that

$$V(x) \leq \int_0^t e^{-s}f(x_n(s),\alpha_n(s))ds + e^{-t}V(x_n(t)) \qquad \text{for } t \geq 0.$$

Thus, using (3.35), (3.36) and Corollary 2.1, we have

$$V(x) < \frac{1}{n} + \int_0^t e^{-s} \sum_{i=1}^\infty \lambda_i(s)f(\xi(s),a_i)ds + e^{-t}V(\xi(t))$$
$$+ e^{-t}(C_1 \frac{1}{n} + 3C_1 + 2\sigma(C_1))$$

for $0 < t \leq T$. Letting $n \to \infty$, we conclude (3.34). Q.E.D.

Proof of Theorem 3.1. For $x \in \mathbb{R}^N$ and $\mu > 0$, we let $\langle x\rangle_\mu$ denote $(|x|^2 + \mu^2)^{1/2}$. We choose a $\theta \in C^\infty(\mathbb{R})$ such that $0 \leqq \theta(r) \leqq r$ for $r \geqq 0$, $\theta(r) = r$ for $r \geqq 1$ and $\theta(r) = 0$ near $r = 0$. For $\nu > 0$, we set $\theta_\nu(r) = \nu\theta(r/\nu)$ for $r \in \mathbb{R}$.

Our goal is to prove that

(3.37) for any $\varepsilon > 0$ there exist $\delta = \delta(\varepsilon) > 0$, $E = E(\varepsilon) > 0$ and $\gamma = \gamma(\varepsilon) \in (0,1)$ such that

$$V(x) - V(y) < \varepsilon + E|x - y|^\gamma \qquad \text{if} \quad |x - y| < \delta.$$

To this end, we fix $\varepsilon > 0$ and choose $C = C(\varepsilon) > 0$ so that

$$\omega(r) \leq \varepsilon + Cr \qquad \text{for} \quad r \geqq 0. \tag{3.38}$$

We define $T = T(\varepsilon) > 0$, $E = E(\varepsilon) > 0$ and $\gamma = \gamma(c) > 0$ by

$$\begin{cases} 2e^{-T}(3C_1 + 2\sigma(C_1)) = \varepsilon, \\ E = \max\{2C,\ 48\varepsilon e^T,\ 32(3C_1 + 2\sigma(C_1)),\ 2C_1\}, \\ \gamma = \min\{\frac{1}{2C}, \frac{1}{2}\}. \end{cases} \tag{3.39}$$

Let $x, y \in \mathbb{R}^N$ satisfy

$$|x - y|^\gamma < \frac{e^{-T}}{16}, \tag{3.40}$$

and choose $\mu > 0$ so that

$$\langle x - y\rangle_\mu^\gamma < \frac{e^{-T}}{16} \tag{3.41}$$

and then $\nu > 0$ so that

$$\nu\gamma\mu^{\gamma-1}e^T < \frac{1}{16} \quad \text{and} \quad \nu\gamma E\mu^{\gamma-1} < \varepsilon. \tag{3.42}$$

We may assume $V(x) \geq V(y)$; otherwise we have nothing to prove. We choose $\beta \in A_{ad}(y)$ and $y(\cdot) \in X_{ad}(y,\beta)$ so that

$$V(y) > -\varepsilon + J(y,\beta,y(\cdot)). \tag{3.43}$$

We apply Proposition 3.2 with

$$p(z,\eta,t) = \gamma E\langle z - y(t)\rangle_{\mu}^{\gamma-1} \frac{\eta - g(y(t),\beta(t))}{|\eta - g(y(t),\beta(t))|^2}\, \theta_{\nu}(|\eta - g(y(t),\beta(t))|),$$

to obtain $\{a_i\}_{i\in\mathbb{N}} \subset A$, $\{\lambda_i\}_{i\in\mathbb{N}} \subset L^{\infty}(0,\infty)$ and an absolutely continuous function $\xi: [0,\infty) \to \mathbb{R}^N$ satisfying (a), (b), (c) and (3.34). Using (A3) and (3.38), we calculate that

$$\begin{aligned}
&\sum_{i=1}^{\infty} \lambda_i(t) f(\xi(t),a_i) - f(y(t),\beta(t)) \\
&< \varepsilon - H(\xi(t),p(\xi(t),\dot{\xi}(t),t)) + H(y(t),p(\xi(t),\dot{\xi}(t),t)) \\
&\qquad + (\dot{y}(t) - \dot{\xi}(t))\cdot p(\xi(t),\dot{\xi}(t),t) \\
&\leqq 2\varepsilon + C|\xi(t) - y(t)|(1 + |p(\xi(t),\dot{\xi}(t),t)|) \\
&\qquad - \gamma E\langle \xi(t) - y(t)\rangle_{\mu}^{\gamma-1}\, \theta_{\nu}(|\dot{\xi}(t) - \dot{y}(t)|) \\
&\leqq 2\varepsilon + C|\xi(t) - y(t)| + \gamma EC\langle \xi(t) - y(t)\rangle_{\mu}^{\gamma} \\
&\qquad - \gamma E\langle \xi(t) - y(t)\rangle_{\mu}^{\gamma-1}\, \theta_{\nu}(|\dot{\xi}(t) - \dot{y}(t)|)
\end{aligned} \tag{3.44}$$

for a. a. $t \geqq 0$.

Next we note that, by (3.34) and (3.43),

$$\begin{aligned}
V(x) - V(y) &< \varepsilon + \int_0^t e^{-s} \{\sum_{i=1}^{\infty} \lambda_i(s) f(\xi(s),a_i) - f(y(s),\beta(s))\} ds \\
&\quad + e^{-t}\{V(\xi(t) - V(y(t)) + 3C_1 + 2\sigma(C_1)\}
\end{aligned}$$

for all $0 < t \leqq T$. Thus, by (3.44) and Corollary 2.1, we have

$$(3.45)\qquad 0 \leq V(x) - V(y) \leq 3\varepsilon + \int_0^t e^{-s}\{C|\xi(s) - y(s)| + \gamma EC\langle\xi(s) - y(s)\rangle_\mu^\gamma$$
$$- \gamma E\langle\xi(s) - y(s)\rangle_\mu^{\gamma-1}\,\theta_\nu(|\dot\xi(s) - \dot y(s)|)\}ds$$
$$+ e^{-t}\{C_1|\xi(t) - y(t)| + 6C_1 + 4\sigma(C_1)\} \qquad \text{for} \quad 0 < t \leq T.$$

Now we want to show that $\langle\xi(t) - y(t)\rangle_\mu < 1$ for $0 \leq t \leq T$. To see this, we suppose that $\langle\xi(t) - y(t)\rangle_\mu \geq 1$ for some $0 \leq t \leq T$, and will obtain a contradiction. In view of (3.41) and the continuity of $t \to \langle\xi(t) - y(t)\rangle_\mu$, there exists $t_0 > 0$ such that

$$\langle\xi(t) - y(t)\rangle_\mu < 1 \quad \text{for} \quad 0 \leq t < t_0 \quad \text{and} \quad \langle\xi(t_0) - y(t_0)\rangle_\mu = 1.$$

Since $C + \gamma EC \leq E$ and

$$\frac{d}{dt}\langle\xi(t) - y(t)\rangle_\mu^\gamma \leq \gamma\langle\xi(t) - y(t)\rangle_\mu^{\gamma-1}(\nu + \theta_\nu(|\dot\xi(t) - \dot y(t)|))$$

for a. a. $t \geq 0$, it follows from (3.45) and an integration by parts that

$$(E - C_1)\langle\xi(t_0) - y(t_0)\rangle_\mu^\gamma e^{-t_0} \leq 3\varepsilon + E\langle x - y\rangle_\mu^\gamma$$
$$+ \nu\gamma E\mu^{\gamma-1} + 2e^{-t_0}(3C_1 + 2\sigma(C_1)).$$

By (3.39), (3.41) and (3.42) we have

$$E - C_1 \geq \frac{E}{2}, \quad 3\varepsilon e^{t_0} \leq \frac{E}{16}, \quad E\langle x - y\rangle_\mu^\gamma e^{t_0} \leq \frac{E}{16},$$
$$\nu\mu E\mu^{\gamma-1} e^{t_0} \leq \frac{E}{16} \quad \text{and} \quad 2(3C_1 + 2\sigma(C_1)) \leq \frac{E}{16}.$$

Therefore

$$\langle\xi(t_0) - y(t_0)\rangle_\mu^\gamma \leq \frac{1}{2},$$

which is a contradiction. Thus we find that $\langle\xi(t) - y(t)\rangle_\mu < 1$ for all $0 \leq t \leq T$.

Using (3.45) with $t = T$, we now find as above that

$$(3.46) \quad V(x) - V(y) \leqq 3\varepsilon + E\langle x - y\rangle_\mu^\gamma + \nu\gamma E\mu^{\gamma-1} + 2e^{-T}(3C_1 + 2\sigma(C_1))$$

$$\leqq 5\varepsilon + E\langle x - y\rangle_\mu^\gamma \quad \text{by (3.42) and (3.39).}$$

After sending $\mu \downarrow 0$, this proves assertion (3.38). We conclude from (3.37) the existence of a function $\delta\colon [0,\infty) \to [0,\infty)$ such that $\delta(0) = 0$, $\delta(r) > 0$ for $r > 0$ and (3.1) holds. Q.E.D.

§4. The Bellman equation

In this section we will prove the following

Theorem 4.1. The value function V is the (unique) viscosity solution of

$$(4.1) \quad \begin{cases} u + H(x,Du) = 0 & \text{in } \mathbf{R}^N, \\ u \in UC(\mathbf{R}^N), \end{cases}$$

where H is defined by (2.1) and satisfies (A1), (A2) and (A3).

We refer the reader to [4, 5] for the definition of viscosity solutions. The uniqueness of the viscosity solution of (4.1) follows from [8]. Such uniqueness results were first obtained in [5] in the class $BUC(\mathbf{R}^N)$.

The basic idea of our proof is the same as that in [11].

Proof. Let $\phi \in C^1(\mathbf{R}^N)$, and assume that $V - \phi$ attains its local minimum at $x_0 \in \mathbf{R}^N$. We will show that

$$(4.2) \quad V(x_0) + H(x_0, D\phi(x_0)) \geqq 0.$$

Since V is uniformly continuous by Theorem 3.1, we may assume that $C_2 \equiv \sup\{|D\phi(x)| \mid x \in \mathbf{R}^N\} < \infty$ and that $V - \phi$ attains its global minimum at x_0.

For any $\varepsilon > 0$ and $t > 0$, by virtue of Proposition 3.1, we can select $\alpha \in A_{ad}(x_0)$ and $x(\cdot) \in X_{ad}(x_0,\alpha)$ so that

$$(4.3) \qquad V(x_0) > -\varepsilon t + \int_0^t e^{-s} f(x(s),\alpha(s))ds + e^{-t}V(x(t)).$$

By the definition of H, we have

$$H(x(s),-(C_1+1)\frac{g(x(s),\alpha(s))}{|g(x(s),\alpha(s))|}) \geqq (C_1+1)|g(x(s),\alpha(s))| - f(x(s),\alpha(s))$$

if $g(x(s),\alpha(s)) \neq 0$, and so

$$f(x(s),\alpha(s)) \geqq |\dot{x}(s)| + C_1 \frac{d}{ds}|x(s)| - H(0,0) - \sigma(C_1+1) - C_1(|x(s)|+1)$$

for a. a. $s \geqq 0$. Plug this into (4.3) and use Theorem 2.1 and integration by parts, to get

$$(4.4) \qquad \int_0^t e^{-s}|\dot{x}(s)|ds < \varepsilon t + C_3$$

for some constant $C_3 > 0$ independent of ε and t.

Let $0 < \delta \leqq 1$ and $R = C_2 + 1$. We see as above that

$$H(x(s),D\phi(x(s)) - \delta\frac{g(x(s),\alpha(s))}{|g(x(s),\alpha(s))|})$$

$$\geqq - g(x(s),\alpha(s))\cdot D\phi(x(s)) + \delta|g(x(s),\alpha(s))| - f(x(s),\alpha(s))$$

if $g(x(s),\alpha(s)) \neq 0$. Hence

$$-f(x(s),\alpha(s)) \leqq H(x(s),D\phi(x(s))) + \sigma_R(\delta) + \frac{d}{ds}(\phi(x(s)) - \delta|x(s) - x_0|)$$

for a. a. $s \geqq 0$. Inserting this into (4.3) and integrating by parts, we obtain

$$V(x_0) - \phi(x_0) - e^{-t}(V(x(t)) - \phi(x(t)))$$

$$+ \int_0^t e^{-s}\{H(x(s),D\phi(x(s))) + \phi(x(s)) - \delta|x(s) - x_0|\}ds$$

$$> - \varepsilon t - \sigma_R(\delta)t + \delta e^{-t}|x(t) - x_0|.$$

Since $V(x) - \phi(x) \geqq V(x_0) - \phi(x_0)$ for $x \in \mathbf{R}^N$, we find that

$$(4.5) \quad (1 - e^{-t})(V(x_0) - \phi(x_0)) + \int_0^t e^{-s}\{H(x(s),D\phi(x(s))) + \phi(x(s))\}ds$$
$$> - \varepsilon t - \sigma_R(\delta)t + \delta e^{-t}|x(t) - x_0|.$$

This inequality, with $\delta = 1$, and (4.4) show that

$$|x(t) - x_0| \leqq C_4 t$$

for all $0 < t \leqq 1$ and some C_4 independent of ε and t. Thus, dividing (4.5) by t and sending $t \downarrow 0$, we have

$$V(x_0) + H(x_0,D\phi(x_0)) > - \varepsilon - \sigma_R(\delta);$$

this implies (4.2).

Now we assume that $V - \phi$ attains its local maximum at $x_0 \in \mathbf{R}^N$ and will prove that

$$(4.6) \qquad V(x_0) + H(x_0,D\phi(x_0)) \leqq 0.$$

To this aim, fix $a \in A$ and choose $\alpha \in A_{ad}(x_0)$ and $x(\cdot) \in X_{ad}(x_0,\alpha)$ so that $\alpha(t) = a$ for $0 \leqq t \leqq 1$. By Proposition 3.1, we have

$$V(x_0) \leqq \int_0^t e^{-s}f(x(s),\alpha(s))ds + e^{-t}V(x(t))$$

for $t > 0$, and hence

$$(1 - e^{-t})(V(x_0) - \phi(x_0)) \leqq \int_0^t e^{-s}f(x(s),\alpha(s))ds + e^{-t}\phi(x(t)) - \phi(x_0)$$

$$= \int_0^t e^{-s}\{f(x(s),\alpha(s)) + D\phi(x(s))\cdot g(x(s),\alpha(s)) - \phi(x(s))\}ds$$

for sufficiently small $t > 0$. Dividing this by t and letting $t \downarrow 0$, we conclude

$$V(x_0) - g(x_0,a)\cdot D\phi(x_0) - f(x_0,a) \leq 0.$$

Since $a \in A$ is arbitrary, we see that (4.6) holds. Q.E.D.

§5. Representation of Hamiltonians

Let $H: \mathbb{R}^N \times \mathbb{R}^N \to \mathbb{R}$ be a given function satisfying (A2), (A3) and the following condition.

(A4) $p \to H(x,p)$ is convex for all $x \in \mathbb{R}^N$.

If necessary, replacing ω by a new one, we may assume in addition:

(5.1) ω is increasing and concave on $[0,\infty)$.

This implies that

$$\omega(r + s) \leq \omega(r) + \omega(s) \qquad \text{for } r, s \geq 0. \tag{5.2}$$

In order to state the main result in this section, we define

$$\begin{aligned} A(\chi,H) = \{(\eta,\zeta) \in C(\mathbb{R}^N) \times C(\mathbb{R}^N, \mathbb{R}^N) \mid\ & |\eta(x) - \eta(y)| \leq L\chi(|x - y|), \\ & |\zeta(x) - \zeta(y)| \leq L\chi(|x - y|) \quad \text{and} \quad -\zeta(x)\cdot p - \eta(x) \leq H(x,p) \\ & \text{for all } x, y, p \in \mathbb{R}^N \text{ and some constant } L > 0\}, \end{aligned}$$

where $\chi: [0,\infty) \to [0,\infty)$.

<u>Theorem 5.1</u>. If (A2), (A3), (A4) and (5.1) hold, then

$$H(x,p) = \max \{-\zeta(x)\cdot p - \eta(x) \mid (\eta,\zeta) \in A(\chi,H)\} \tag{5.3}$$

for the function $\chi: r \to \sqrt{\omega(r)} + \omega(r)$.

To prove Theorem 5.1, we need the next selection lemma.

Lemma 5.1. Let $K: \mathbb{R}^N \to P(\mathbb{R}^M)$, and assume that $K(x)$ is nonempty, closed and convex for $x \in \mathbb{R}^N$. Assume further that

$$K(x) \subset K(y) + B^M(0,\chi(|x - y|)), \tag{5.4}$$

$$K(x) \cap B^M(0,R) \neq \emptyset \tag{5.5}$$

for all $x, y \in \mathbb{R}^N$ and some constant $R > 0$, where $\chi: [0,\infty) \to [0,\infty)$ is an increasing, concave, continuous function satisfying $\chi(0) = 0$. Let $x_0 \in \mathbb{R}^N$ and $\xi_0 \in \mathbb{R}^M$ satisfy $\xi_0 \in K(x_0)$. Then there exist a function $\xi: \mathbb{R}^N \to \mathbb{R}^M$ and a constant $L > 0$ depending on ξ_0 and R such that

$$\xi(x_0) = \xi_0, \tag{5.6}$$

$$\xi(x) \in K(x), \tag{5.7}$$

$$|\xi(x) - \xi(y)|^2 \leqq L\chi(|x - y|) \tag{5.8}$$

for $x, y \in \mathbb{R}^N$.

Proof. For any $x \in \mathbb{R}^N$, taking into account that $K(x)$ is a nonempty, closed, convex subset of $\mathbb{R}^M$, we set

$$\xi(x) = \text{the nearest point of } K(x) \text{ from } \xi_0.$$

It is obvious that (5.6) and (5.7) hold.

To see (5.8), we first note that $|\xi(x) - \xi_0| \leqq |\xi_0| + R$ for $x \in \mathbb{R}^N$ by (5.5). Let $x, y \in \mathbb{R}^N$ and assume $|\xi(x) - \xi_0| \geqq |\xi(y) - \xi_0|$ for definiteness. We may assume $|\xi(x) - \xi_0| > 0$. Let $\eta \in \mathbb{R}^M$ be the nearest point from $\xi(y)$ on the line determined by two points $\xi(x)$ and ξ_0, i. e., $\eta = \xi_0 + ((\xi(y) - \xi_0)\cdot v)v$, where $v = (\xi(x) - \xi_0)/|\xi(x) - \xi_0|$. A simple calculation or plane geometry shows that

$$|\xi(x) - \xi(y)|^2 = |\xi(y) - \xi_0|^2 - |\xi(x) - \xi_0|^2 + 2|\xi(x) - \xi_0||\xi(x) - \eta|$$

$$\leq 2|\xi(x) - \xi_0||\xi(x) - \eta|.$$

Since $|\xi(x) - \eta| \leq \chi(|x - y|)$ by (5.4), we conclude that

$$|\xi(x) - \xi(y)|^2 \leq 2(|\xi_0| + R)\chi(|x - y|).$$

This completes the proof. Q.E.D.

Proof of Theorem 5.1. Let $\tilde{\chi}$ be the function on $[0,\infty)$ defined by $\tilde{\chi}(r) = \sqrt{\omega(r)}$. Define $H_1: \mathbb{R}^N \times \mathbb{R}^N \to \mathbb{R}$ by $H_1(x,p) = H(x,p) - H(x,0)$ for x, $p \in \mathbb{R}^N$. We will prove

$$(5.9) \qquad H_1(x,p) = \max \{- \zeta(x)\cdot p - \eta(x) | (\eta,\zeta) \in A(\tilde{\chi},H_1)\}$$

for x, $p \in \mathbb{R}^N$. It is clear that (5.3) holds.

By the definition of $A(\tilde{\chi},H_1)$, we have

$$H_1(x,p) \geq \sup \{- \zeta(x)\cdot p - \eta(x) | (\eta,\zeta) \in A(\tilde{\chi},H_1)\}$$

for x, $p \in \mathbb{R}^N$. Therefore it is enough to see that there exists $(\eta,\zeta) \in A(\tilde{\chi},H_1)$ for each x, $p \in \mathbb{R}^N$ such that

$$(5.10) \qquad H_1(x,p) = - \zeta(x)\cdot p - \eta(x).$$

We define $K: \mathbb{R}^N \to P(\mathbb{R}^{N+1})$ by setting

$$K(x) = \{ (\eta,\zeta) \in \mathbb{R} \times \mathbb{R}^N | - \zeta\cdot p - \eta \leq H_1(x,p) \quad \text{for} \quad p \in \mathbb{R}^N\}$$

for $x \in \mathbb{R}^N$. It is easy to check that $K(x)$ is nonempty, closed and convex for all $x \in \mathbb{R}^N$.

We claim that

$$\text{(5.11)} \qquad K(x) \subset K(y) + B^{N+1}(0, 2\sqrt{2}\omega(|x - y|))$$

for $x, y \in \mathbb{R}^N$. To see this, fix $x \neq y \in \mathbb{R}^N$, and choose $\varepsilon \geqq 0$ and $C_\varepsilon > 0$ so that

$$\text{(5.12)} \qquad \begin{cases} 2\omega(r) \leqq \varepsilon + C_\varepsilon r & \text{for all } r \geqq 0 \text{ and} \\ 2\omega(r) = \varepsilon + C_\varepsilon |x - y|. \end{cases}$$

Let $(\eta_x, \zeta_x) \in K(x)$, and observe that

$$H_1(y,p) \geqq H_1(x,p) - 2\omega(|x - y|(1 + |p|))$$

$$\geqq - \zeta_x \cdot p - \eta_x - \varepsilon - C_\varepsilon |x - y|(1 + |p|)$$

for $p \in \mathbb{R}^N$ by (A3) and (5.12). Note that

$$p \to - \zeta_x \cdot p - \eta_x - \varepsilon - C_\varepsilon |x - y|(1 + |p|)$$

is concave on $\mathbb{R}^N$. Therefore, by a standard separation theorem, there exists $(\eta_y, \zeta_y) \in \mathbb{R} \times \mathbb{R}^N$ such that

$$H_1(y,p) \geqq - \zeta_y \cdot p - \eta_y \geqq - \zeta_x \cdot p - \eta_x - \varepsilon - C_\varepsilon |x - y|(1 + |p|)$$

for $p \in \mathbb{R}^N$. From this we see that

$$(\eta_y, \zeta_y) \in K(y)$$

and

$$(\zeta_y - \zeta_x) \cdot p + \eta_y - \eta_x \leqq \varepsilon + C_\varepsilon |x - y|(1 + |p|)$$

for $p \in \mathbb{R}^N$. Inserting $p = 0$ into this inequality, we have

(5.13) $$\eta_y - \eta_x \leqq \varepsilon + C_\varepsilon|x - y| = 2\omega(|x - y|)$$

and taking $p = t(\zeta_y - \zeta_x)/|\zeta_y - \zeta_x|$, dividing by t and sending $t \to \infty$ if $\zeta_y \neq \zeta_x$, we have also

(5.14) $$|\zeta_y - \zeta_x| \leqq C_\varepsilon|x - y| \leqq 2\omega(|x - y|).$$

Note that if $(\eta,\zeta) \in K(y)$ and $\eta < \eta'$, then $(\eta',\zeta) \in K(y)$. Thus, replacing η_y by a larger number if necessary, we conclude that there exists $(\eta_y,\zeta_y) \in K(y)$ such that (5.14) and

$$|\eta_y - \eta_x| \leqq 2\omega(|x - y|)$$

hold. We have proved (5.11).

Next we show that

(5.15) $$K(x) \cap B^{N+1}(0,R) \neq \emptyset$$

for $x \in \mathbb{R}^N$ and some constant $R > 0$. Since $H_1(x,0) = 0$ for $x \in \mathbb{R}^N$, by a separation theorem there exists a $\zeta_x \in \mathbb{R}^N$ for any $x \in \mathbb{R}^N$ such that $-\zeta_x \cdot p \leqq H_1(x,p)$ for $p \in \mathbb{R}^N$. If $\zeta_x \neq 0$, then

$$|\zeta_x| \leqq H_1(x,-\zeta_x/|\zeta_x|) \leqq \sigma(1) \qquad \text{by (A2).}$$

Thus $(0,\zeta_x) \in K(x) \cap B^{N+1}(0,\sigma(1))$ for $x \in \mathbb{R}^N$, which proves (5.15).

Fix x_0, $p_0 \in \mathbb{R}^N$, and choose $(\eta_0,\zeta_0) \in K(x_0)$ so that $H_1(x_0,p_0) = -\zeta_0 \cdot p_0 - \eta_0$. By Lemma 5.1, there exist $\eta\colon \mathbb{R}^N \to \mathbb{R}$, $\zeta\colon \mathbb{R}^N \to \mathbb{R}^N$ and $L > 0$ such that $\eta(x_0) = \eta_0$, $\zeta(x_0) = \zeta_0$,

$$(\eta(x),\zeta(x)) \in K(x) \qquad \text{for } x \in \mathbb{R}^N,$$

$$|\eta(x) - \eta(y)| \leqq L\tilde{\chi}(|x - y|) \qquad \text{for } x, y \in \mathbb{R}^N,$$

$$|\zeta(x) - \zeta(y)| \leq L\tilde{\chi}(|x - y|) \qquad \text{for} \quad x, y \in \mathbb{R}^N.$$

The proof is now finished.

§6. Representation of viscosity solutions

As in the previous section we assume here that a function $H: \mathbb{R}^N \times \mathbb{R}^N \to \mathbb{R}^N$ is given and satisfies (A2), (A3) and (A4). We assume also (5.1). Our purpose is to represent the viscosity solution of

$$(6.1) \qquad \begin{cases} \dot{u} + H(x, Du) = 0 & \text{in } \mathbb{R}^N, \\ u \in UC(\mathbb{R}^N) \end{cases}$$

as the value function of an appropriate optimal control problem.

By Theorem 5.1, we have

$$(6.2) \qquad H(x,p) = \max \{- \zeta(x)\cdot p - \eta(x) \mid (\eta,\zeta) \in A(\chi,H)$$

for $x, p \in \mathbb{R}^N$, where χ is the function: $r \to \sqrt{\omega(r)} + \omega(r)$ on $[0,\infty)$.

We reard $A = A(\chi,H)$ as the metric soace equipped with the usual metric in $C(\mathbb{R}^N, \mathbb{R}^{N+1})$. We define $f: \mathbb{R}^N \times A \to \mathbb{R}$ and $g: \mathbb{R}^N \times A \to \mathbb{R}$ by setting

$$f(x,a) = \eta(x) \qquad \text{and} \qquad g(x,a) = \zeta(x)$$

for $a = (\eta,\zeta) \in A(\chi,H)$. It is clear that f and g satisfy (A1). It is immediate from (6.2) that

$$(6.3) \qquad H(x,p) = \max_{a \in A} \{- g(x,a)\cdot p - f(x,a)\} \qquad \text{for} \quad x, p \in \mathbb{R}^N.$$

Thus, in view of the uniqueness of the viscosity solution of (6.1), we have the following theorem as a consequence of Theorem 4.1.

Theorem 6.1. The viscosity solution u of (6.1) is represented as

$$u(x) = V(x) \equiv \inf \{\int_0^\infty e^{-t} f(x(t),\alpha(t))dt \mid \alpha \in A_{ad}(x),\ x(\cdot) \in X_{ad}(x,\alpha)\}. \tag{6.4}$$

Remark 6.1. This theorem extends a result of Lions [11]. Evans-Souganidis [6] treated the same representation problem without assuming (A4) but requiring H to be represented as a "max-min" of $-g(x,a,b)\cdot p - f(x,a,b)$, where f and g satisfy some Lipschitz continuity assumption.

References

1. G. Barles, Thése de 3e cycle, Univ. Paris IX - Dauphine, Paris, 1982/1983.

2. N. E. Barron, L. C. Evans and R. Jensen, Viscosity solutions of Isaacs' equations and differential games with Lipschitz controls, to appear in J. Diff. Eq.

3. F. H. Clarke, Optimization and Nonsmooth Analysis, Wiley, New York, 1983.

4. M. G. Crandall, L. C. Evans and P. L. Lions, Some properties of viscosity solutions of Hamliton-Jacobi equations, Trans. AMS. 282 (1984), 487 - 502.

5. M. G. Crandall and P. L. Lions, Viscosity solutions of Hamilton-Jacobi equations, Trans. AMS. 277 (1983), 1 - 42.

6. L. C. Evans and P. E. Souganidis, Differential games and representation formulas for solutions of Hamilton-Jacobi-Isaacs equations, to appear in Indiana Univ. Math. J.

7. W. H. Fleming and R. W. Rishel, Deterministic and Stochastic Optimal Control, Springer-Verlag, New York, 1975.

8. H. Ishii, Uniqueness of unbounded viscosity solutions of Hamilton-Jacobi equations, to appear in Indiana Univ. Math. J.

9. H. Ishii, Remarks on existence of viscosity solutions of Hamilton-Jacobi equations, Bull. Facul. Sci. & Eng. Chuo Univ. 26 (1983), 5 - 24.

10. S. Kakutani, A generalization of Brouwer's fixed point theorem, Duke Math. J., 8 (1941), 457 - 459.

11. P. L. Lions, Generalized Solutions of Hamilton-Jacobi Equations, Pitman, Boston, 1982.

12. P. L. Lions, Existence results for first-order Hamilton-Jacobi equations, Ric. Mat. Napoli, 32 (1983), 3 - 23.

13. E. Michael, Continuous selections. I, Ann. Math., 63 (1956), 361 - 382.

14. P. E. Souganidis, Thesis, Univ. of Wisconsin, 1983.

15. P. E. Souganidis, Existence of viscosity solutions of Hamilton-Jacobi equations, to appear in J. Diff. Eq.

16. J. Warga, Optimal Control of Differential and Functional Equations, Academic Press, New York, 1972.

Lecture Notes in Num. Appl. Anal., 8, 53–71 (1985)
Recent Topics in Nonlinear PDE II, Sendai, 1985

The Existence of Nonstationary Ideal Incompressible Flow in Exterior Domains in R^3

Keisuke Kikuchi

Department of Pure and Applied Sciences
College of General Education
Tokyo 153, Japan

Introduction.

We consider the motion of an ideal incompressible fluid past a finite number of isolated rigid bodies $O_1,\dots,O_m$ in R^3. The velocity $v = (v^1(x,t),v^2(x,t),v^3(x,t))$ and the (scalar) pressure $p = p(x,t)$ of the fluid motion are governed by the Euler equation in $\Omega = R^3\setminus(O_1\cup\dots\cup O_m)$

$$(1)\qquad \frac{\partial v}{\partial t} + (v\cdot\nabla)v + \nabla p = f, \qquad \operatorname{div} v = 0 \qquad t\in[0,T]$$

subject to the following conditions at infinity and on the boundary $S = \partial\Omega$

$$(2)\qquad \lim_{|x|\to\infty} v(x,t) = v_\infty, \qquad v\cdot n_{|S} = 0 \qquad t\in[0,T]$$

satisfying the initial condition

$$(3)\qquad v(x,t) = v_0(x),$$

where $f = f(x,t)$ is a given external force vector, $v_0(x)$ is a given initial velocity and v_∞ is a given constant vector.

The purpose of this paper is to give the outline of the proof of the existence of a solution $\{v, p\}$ of (1), (2) and (3) which satisfies the asymptotic condition at infinity that v converges to v_∞ faster than $|x|^{-(1+\delta)}$ for a certain $\delta \geqq 0$. For further details, see [16].

The problem of existence and uniqueness of solutions of the Euler equation has been considered by several authors. Recently, when $\Omega = R^3$, this problem was studied by Swann [18], Kato [13], Bardos and Frisch [2] and Cantor [5]. When Ω is bounded in R^3, the problem was studied by Ebin and Marsden [8], Swann [19], Bourguignon and Brézis [4], Temam [20] and Kato and Lai [14]. In the two-dimensional case, the existence of a global solution was studied by Judovič [11] and Kato [12] (in a bounded case), and by Kikuchi [15] (in an unbounded case). Among the above papers, the works of Cantor [5] and Swann [19] especially inspire the idea of our proof. In [5], Cantor constructed solutions introducing the weighted Sobolev space over R^3, on which the Laplacian is an isomorphism (see Lemma 2.4 in this paper). This work suggested to the author to use the corresponding space on the exterior domains. The advantage to the property of the weighted Sobolev space makes possible the construction of solutions of linear elliptic system (see Proposition 3.1). Our proof is essentially a modified form of that of Swann [19], who constructed solutions by considering the vorticity equation, with suitable treatment of the exterior problem (see Proposition 3.2).

I. Notations and statement of the result.

The domain Ω is simply connected and $R^3\setminus\Omega$ consists of m numbers of compact components $O_1,\dots,O_m$; the boundary $S = S_1 + \dots + S_m$ ($S_j = \partial O_j$) is sufficiently smooth.

For $1 \leq p \leq \infty$, the norm in $L^p(\Omega)$ is denoted by $|\cdot|_p$. For $s \geq 0$ (integer) and $1 \leq p \leq \infty$, the norm in $W^{s,p}(\Omega)$ is denoted by $|\cdot|_{s,p}$. For $s \geq 0$ (integer), $C_b^s(\bar\Omega)$ is the set of all $u \in C^s(\bar\Omega)$ such that all their derivatives up to order s are bounded. $C_0^s(\bar\Omega)$, s being non-negative integer, is the set of all functions that belong to $C^s(\bar\Omega)$ and have compact support in $\bar\Omega$.

Let $\sigma(x) = (1+|x|^2)^{1/2}$. For $p \geq 1$, $\lambda \geq 0$ and $s \geq 0$, $M^p_{s,\lambda}$ denotes the closure of $C_0^s(\bar\Omega)$ with respect to the norm

$$|u|_{p,s,\lambda} = \sum_{|\alpha|\leq s} |\sigma^{\lambda+|\alpha|} D^\alpha u|_p .$$

We write

$$\|u\|_{p,0,\lambda} = \sup_{t\in[0,T]} |u(.,t)|_{p,s,\lambda},$$

$$\|u\|_{s,p} = \sup_{t\in[0,T]} |u(.,t)|_{s,p} \qquad (|\cdot|_{0,p} = |\cdot|_p).$$

For a vector-valued function u, $u\cdot n_{|S}$ is the outward normal component of u on S, and we write

$$Du = \left(\frac{\partial u}{\partial x_j}\ ;\ j = 1,2,3\right), \qquad D^2u = \left(\frac{\partial^2 u}{\partial x_j \partial x_k};\ j,k = 1,2,3\right).$$

Now we can state our main result.

Theorem 1.1. Let $p > 3$ and $0 \leq \delta < 1-3/p$. Assume that

(i) v_0 belongs to $C_b(\bar{\Omega}) \cap C^1(\Omega)$, rot $v_0 \in M^p_{1,\delta+2}$;

$\operatorname{div} v_0 = 0$; $v_0 \cdot n_{|S} = 0$; $\lim_{|x|\to\infty} v_0 = v_\infty$,

(ii) f belongs to $C([0,T];C_b(\bar{\Omega}) \cap C^1(\Omega))$,

rot $f \in L^\infty((0,T);M^p_{1,\delta+2})$.

Then there exists $T_0 > 0$, $T_0 \leq T$, depending only on rot v_0, v_∞, rot f and Ω, such that (1), (2) and (3) have a solution $\{v, p\}$ on $[0,T_0]$ satisfying

$$v - v_\infty \in C([0,T_0];M^p_{1,\delta+1}) \cap L^\infty((0,T_0);M^p_{2,\delta+1}),$$

(1.1)

$$\frac{\partial v}{\partial t} \in L^\infty((0,T_0);M^p_{1,\delta+1}), \quad \nabla p \in L^\infty((0,T_0);C_b(\bar{\Omega})).$$

Such a solution is unique up to an arbitrary function of t which may be added to p.

II. Preliminaries.

2.1. Properties of $M^p_{s,\delta}$.

The Hölder inequality implies

Lemma 2.1. Let $\lambda \geq 0$. If $\sigma^\lambda u \in L^p(\Omega)$, then $u \in L^r(\Omega)$ provided that $3p/(\lambda p+3) < r \leq p$.

Lemma 2.2. (The Sobolev imbedding theorem.) Let $s \geq 1$, $p > 3$ and $\lambda \geq 0$. Then $M^p_{s,\lambda} \subset C^{s-1}_b(\bar{\Omega})$ and

(2.1) $$|u|_{s-1,\infty} \leq c_1 |u|_{s,p} \leq c_1 |u|_{p,s,\lambda}.$$

In particular, if $\lambda > 0$, then

(2.2) $|u(x)| = O(|x|^{-\lambda})$ as $|x| \to \infty$.

Proof. $|u(x)| = \sigma(x)^{-\lambda}|\sigma^{\lambda}u|_{\infty} \leq c_1(1+|x|^2)^{-\lambda/2}|u|_{p,s,\lambda}$.

Lemma 2.3. (Cantor [5]: Proposition 1.1.) If $p > 1$, $s > 3/p$, $\delta \geq 0$ and $0 \leq k \leq s$, then the pointwise multiplication of functions:

$$M^p_{s,\delta} \times M^p_{s-k,\delta+k} \ni (\varphi,\psi) \longmapsto \varphi\cdot\psi \in M^p_{s-k,\delta+k}$$

induces a continuous map.

Lemma 2.4. (Cantor [6]: Theorem 2. Also see Nirenberg and Walker [17]: Theorem 2.1.) Let $p > 3$, $s \geq 0$ and $3/p-3 < \delta < 1-3/p$. Then the Laplace operator

$$\Delta : M^p_{s+2,\delta}(R^3) \longrightarrow M^p_{s,\delta+2}(R^3)$$

is an isomorphism.

Lemma 2.5. (Cantor [7]: Theorem 2.1.) Let $p > 3$, $s \geq 0$ and $0 \leq \delta < 1-3/p$. Then the map N defined by $N(u) = (-\Delta u, \frac{\partial u}{\partial n})$

$$N: M^p_{s+2,\delta} \longrightarrow M^p_{s,\delta+2} \times W^{s+1-1/p,p}(S)$$

is an isomorphism.

2.2. Some boundary value problem.

Lemma 2.6. Let $u \in C(\Omega)$ be a vector function such that rot $u = 0$ (generalized). Then there is a scalar function $q \in C^1(\Omega)$ such that $u = \nabla q$. If, in addition, div $u = 0$ (generalized) and $\lim_{|x|\to\infty} u(x) = 0$, then q is harmonic and satisfies $\lim_{|x|\to\infty} q(x) = \text{const}$.

Proof. Since Ω is simply connected, as is well known, if $u \in C^1(\Omega)$, then rot $u = 0$ implies that the following equation is well-defined:

$$(2.3) \qquad q(x) = \int_{x_0}^{x} u(y)\cdot\tau(y)d_y\Gamma + q(x_0),$$

where the integral of u is along any path in Ω from a fixed point x_0 to x, and $q(x_0)$ is an arbitrarily given constant. Then this q has the required properties (see [16]).

Furthermore, the uniqueness theorem for the exterior Neumann problem implies

Corollary 2.7. Let u be a harmonic field (rot $u = 0$ and div $u = 0$) satisfying $u\cdot n_{|S} = 0$ and tending to 0 at infinity. Then $u = 0$.

III. Construction of solutions.

In this section we shall prove Theorem 1.1. To this end we consider the vorticity equation obtained by taking the rotation of the first equation of (1) and using div $v = 0$:

$$(3.1) \qquad \text{rot } v = w,$$

$$(3.2) \qquad \text{div } v = 0,$$

$$(3.3) \qquad \frac{\partial w}{\partial t} + (v\cdot\nabla)w - (w\cdot\nabla)v = \text{rot } f$$

with the boundary conditions (2) for v and the initial condition:

$$(3.4) \qquad w(x,0) = \text{rot } v_0(x)$$

for w. We construct solutions of the vorticity equation by means of the following iterative process. The vectors $v_0(x)$ and $w_0(x) = \text{rot } v_0(x)$, which are the initial velocity and the initial vorticity respectively, are taken as the zeroth approximations. When the n-th approximation for the vorticity $w_n(x,t)$ is known, then the n-th approximation for the velocity $v_n(x,t)$ is determined as follows:

$$\text{(3.5)} \quad \text{rot } v_n = w_n, \quad \text{div } v_n = 0, \quad v_n \cdot n_{|S} = 0, \quad \lim_{|x|\to\infty} v_n = v_\infty.$$

(For the zeroth approximation (3.5) is automatically satisfied.) And when the n-th approximation for the velocity $v_n(x,t)$ is known, the (n+1)-th approximation for the vorticity $w_{n+1}(x,t)$ is a solution of the following equations:

$$\text{(3.6)} \quad \begin{aligned} &\frac{\partial}{\partial t} w_{n+1} + (v_n \cdot \nabla) w_{n+1} - (w_{n+1} \cdot \nabla) v_n = \text{rot } f, \\ &\text{div } w_{n+1} = 0, \qquad w_{n+1}(x,0) = \text{rot } v_0(x). \end{aligned}$$

The following Propositions 3.1 and 3.2 imply that (3.5) and (3.6) are solvable for all n ($n = 0,1,2,\ldots$) and Lemma 3.3 gives estimates for w_n and v_n which are uniform in n.

We need

<u>Definition.</u> Let p and δ be as in Theorem 1.1 and $s \geq 1$. We define

$$Y^p_{1,\delta+2} = \{w \in M^p_{1,\delta+2}: \ \text{div } w = 0, \ \int_{S_j} (w \cdot n) dS = 0 \ (j = 1,\ldots,m)\}$$

$$X^p_{s,\delta+1} = \{v \in M^p_{s,\delta+1}: \ \text{div } v = 0, \ v \cdot n_{|S} = 0\}.$$

Proposition 3.1. Let $w \in C([0,T];Y^p_{1,\delta+2})$. Then there is a unique solution v of (3.1) and (3.2) under the boundary condition (2). This v satisfies $v-v_\infty \in C([0,T];M^p_{2,\delta+1})$. In addition, there are constants $c_2 = c_2(\Omega)$ and $c_3 = c_3(\Omega, v_\infty)$ such that

(3.7) $$\|v-v_\infty\|_{p,2,\delta+1}$$

$$\leqq c_2(\|w\|_{p,1,\delta+2}+|\mathrm{rot}\ v_0|_{p,1,\delta+2}+c_3)+|v_0-v_\infty|_{p,2,\delta+1},$$

(3.8) $$\|v\|_{1,\infty}+\|D^2v\|_{p,0,\delta+3} \leqq c_2(\|w\|_{p,1,\delta+2}+|\mathrm{rot}\ v_0|_{p,1,\delta+2}+c_3).$$

Proposition 3.2. Let $v \in v_\infty+C([0,T];M^p_{2,\delta+1})$ be such that $\mathrm{div}\ v = 0$ and $v\cdot n_{|S} = 0$. Then there is a unique solution $w \in C([0,T];Y^p_{1,\delta+2})$ of (3.3) and (3.4) satisfying

(3.9) $$|w(t)|_{p,0,\delta+2} \leqq c_4\{1+(T\|v\|_\infty)^{\delta+2}\}(|\mathrm{rot}\ v_0|_{p,0,\delta+2}$$

$$+\int_0^t |\mathrm{rot}\ f(s)|_{p,0,\delta+2}ds + \|Dv\|_\infty\int_0^t |w(s)|_{p,0,\delta+2}ds) \quad t \in [0,T],$$

where the constant c_4 depends only on δ. This solution also satisfies

(3.10) $$|Dw(t)|_{p,0,\delta+3}$$

$$\leqq c_4 e^{T\|Dv\|_\infty}\{1+(T\|v\|_\infty)^{\delta+3}\}(|D(\mathrm{rot}\ v_0)|_{p,0,\delta+3}$$

$$+\int_0^t |D(\mathrm{rot}\ f(s))|_{p,0,\delta+3}ds + \|Dv\|_\infty\int_0^t |Dw(s)|_{p,0,\delta+3}ds$$

$$+ c_1 \sup_{s\in[0,t]} |w(s)|_{p,1,\delta+2}\int_0^t |D^2v(s)|_{p,0,\delta+3}ds) \qquad t \in [0,T].$$

Lemma 3.3. There exist positive constants K_1, K_2 and T_1 ($T_1 \leq T$) depending only on rot v_0, v_∞, rot f and Ω and satisfying

$$(3.11) \qquad \sup_{t\in[0,T_1]} |w_n(t)|_{p,1,\delta+2} \leq K_1,$$

$$(3.12) \qquad \sup_{t\in[0,T_1]} |v_n(t)-v_\infty|_{p,2,\delta+1} \leq K_2 \qquad \text{for all } n.$$

3.1. Sketch of the proof of Proposition 3.1.

The following estimate holds ([16]).

Lemma 3.4. There is a positive constant $c_5 = c_5(\Omega, s)$ such that

$$(3.13) \qquad |v|_{p,s,\delta+1} \leq c_5 |\text{rot } v|_{p,s-1,\delta+2} \qquad \text{for all } v \in X^p_{s,\delta+1}.$$

Lemma 3.5. Let $w \in Y^p_{1,\delta+2}$. Then there is a unique solution $v \in X^p_{2,\delta+1}$ of

$$(3.14) \qquad \text{rot } v = w.$$

Proof. The uniqueness follows from (2.2) and Corollary 2.7. To prove the existence, we need to show that w can be extended to a solenoidal vector function $\tilde{w} \in M^p_{1,\delta+2}(R^3)$.

We construct vector functions w_j ($j = 1,\ldots,m$) such that

$$(3.15) \qquad w_j \in W^{1,p}(O_j); \quad \text{div } w_j = 0; \quad w_j|_{S_j} = w|_{S_j}.$$

For each j, there is a solution $q_j \in W^{2,p}(O_j)$ of the Neumann problem:

$$(3.16) \qquad -\Delta q_j = 0 \quad \text{in } O_j, \quad \frac{\partial q_j}{\partial n} = (w|_{S_j})\cdot n \quad \text{on } S_j$$

(see Agmon, Douglis and Nirenberg [1]). Using the inverse theorem on traces (see Besov, Il'in and Nikol'skii [3: Theorem 25.2]), we have $u_j \in W^{2,p}(O_j)$ $(j = 1,\ldots,m)$ satisfying the boundary condition:

$$(3.17) \qquad u_j|_{S_j} = 0, \quad \frac{\partial u_j}{\partial n}\Big|_{S_j} = w|_{S_j} - (\nabla q_j)|_{S_j}.$$

Furthermore, there exist smooth scalar functions φ_j such that

$$(3.18) \qquad \nabla\varphi_j = n \qquad \text{on } S_j \quad (j = 1,\ldots,m).$$

We put

$$(3.19) \qquad w_j = \mathrm{rot}(\nabla\varphi_j \wedge u_j) + \nabla q_j.$$

Then we have

$$w_j = (\nabla\varphi_j\cdot\nabla)u_j - (u_j\cdot\nabla)\nabla\varphi_j + (\mathrm{div}\, u_j)\nabla\varphi_j - (\Delta\varphi_j)u_j + \nabla q_j.$$

Hence since it follows from (3.16) and (3.17) that $\mathrm{div}\, u_j = 0$ on S_j, we can easily see that (3.15) holds. (The above construction of w_j is suggested by Itô [10: Lemma 2.9].)

Now we define a vector function $\tilde{w}$ on R^3 by

$$(3.20) \qquad \tilde{w} = \begin{cases} w & \text{in } \Omega \\ w_j & \text{on } O_j \quad (j = 1,\ldots,m). \end{cases}$$

Then (3.15) implies that $\tilde{w} \in M^p_{1,\delta+2}(R^3)$ and $\mathrm{div}\,\tilde{w} = 0$. From Lemma 2.4, it follows that there is a solution $u \in M^p_{3,\delta}(R^3)$ of $-\Delta u = \tilde{w}$. In addition, we can see that $\mathrm{div}\, u = 0$ since $\tilde{w}$ is solenoidal. Writing $\bar{v} = \mathrm{rot}\,(u|_\Omega)$, we have

$$(3.21) \qquad \bar{v} \in M^p_{2,\delta+1}; \quad \mathrm{rot}\,\bar{v} = w; \quad \mathrm{div}\,\bar{v} = 0.$$

Lemma 2.5 yields that there is a solution $q \in M^p_{3,\delta}$ of the Neumann problem:

$$(3.22) \qquad -\Delta q = 0, \qquad \frac{\partial q}{\partial n}\Big|_S = \overline{v}\cdot n_{|S}.$$

We put

$$(3.23) \qquad v = \overline{v} - \nabla q.$$

Then it follows from (3.21) and (3.22) that $v \in X^p_{2,\delta+1}$ is a solution of (3.14).

<u>Definition.</u> Let $w \in Y^p_{1,\delta+2}$ and let $v \in X^p_{2,\delta+1}$ be the solution of (3.14). We define the operator F as

$$(3.24) \qquad F(w) = v.$$

Then Lemma 3.4 implies that

$$(3.25) \qquad |F(w)|_{p,2,\delta+1} \leqq c_5 |w|_{p,1,\delta+2}.$$

Furthermore, we have

<u>Corollary 3.6.</u> F is continuous from $C([0,T];Y^p_{1,\delta+2})$ to $C([0,T];X^p_{2,\delta+1})$ with

$$(3.26) \qquad \|F(w)\|_{p,2,\delta+1} \leqq c_5 \|w\|_{p,1,\delta+2}$$

for any $w \in C([0,T];Y^p_{1,\delta+2})$.

It follows from Lemma 3.5 and Corollaries 2.7, 3.6 that

$$(3.27) \qquad v = F(w) + v_0 - F(\mathrm{rot}\ v_0)$$

gives a unique solution of (3.1), (3.2) and (2). We can

prove that this v satisfies (3.7) and (3.8) ([16]).

3.2. Sketch of the proof of Proposition 3.2.

We construct a solution of (3.6) following Swann's argument (see [19], also see Kato [12] and Judovič [11]).

Define a family of curves $(X(x,t;s),s)$ in $\bar{\Omega}\times[0,T]$ by

$$(3.28)\qquad \begin{cases} \dfrac{d}{ds}X(x,t;s) = v(X(x,t;s),s) & t,s\in[0,T] \\ X(x,t;t) = x. \end{cases}$$

Then since $v-v_\infty\in C([0,T];M^p_{2,\delta+1})\subset C([0,T];C^{1+\theta}(\bar{\Omega}))$,

$$(3.29)\qquad X(x,t;s)\in C^1([0,T];C^{1+\theta}(\bar{\Omega}))\qquad (0<\theta<1-3/p).$$

Hence (3.28) gives a unique local curve in $\bar{\Omega}\times[0,T]$ for each $(x,t)\in\bar{\Omega}\times[0,T]$. It follows from $v\cdot n_{|S}=0$ that

$$(3.30)\qquad (X(x,t;s),s) \text{ in } \Omega\times[0,T] \text{ cannot reach } S\times[0,T].$$

Hence all solutions of (3.28) in $\bar{\Omega}\times[0,T]$ exist globally (see Kato [12: Lemma 2.2]). Furthermore the definition (3.28) of stream lines implies that

$$(3.31)\qquad X(X(x,t;s),s;\tau) = X(x,t;\tau)\qquad s,t,\tau\in[0,T].$$

We write the Jacobi matrix of $X(x,t;s)$ as

$$(3.32)\qquad G(x,t;s) = \left(\frac{\partial X^k}{\partial x_j}(x,t;s)\right)_{j,k=1,2,3}.$$

Then we easily see that

$$(3.33)\qquad \det G(x,t;s) = 1.$$

Indeed, differentiating (3.28) in x_j, we see that

$$(3.34)\qquad G(x,t;t) = E \quad \text{(identity matrix)}$$

and that $G(x,t;s)$ is a fundamental matrix solution of the homogeneous ordinary differential equation:

$$(3.35)\qquad \frac{d}{ds} z(s) = (z(s)\cdot\nabla_y)v(y,s)|_{y = X(x,t;s)}.$$

The trace of the coefficient matrix of this equation is $[\operatorname{div} v(y,s)]|_{y=X(x,t;s)}$. Hence from div $v = 0$ we deduce (3.33).

We define

$$(3.36)\qquad w(x,t) = G(x,t;0)^{-1}a(X(x,t;0)) + \int_0^t G(x,t;s)^{-1}b(X(x,t;s),s)ds,$$

where $a = \operatorname{rot} v_0$ and $b = \operatorname{rot} f$. Then using (3.28)$\sim$(3.35) we can prove that w given by (3.36) satisfies the required properties ([16]).

3.3. Proof of Lemma 3.3.

Let

$$(3.37)\qquad K_1 = 4ec_4(|\operatorname{rot} v_0|_{p,1,\delta+2} + \int_0^T |\operatorname{rot} f(t)|_{p,1,\delta+2}dt,$$

$$(3.38)\qquad T_1 = \min[T, 1/\{4e(c_1+1)c_2c_4(K_1+|\operatorname{rot} v_0|_{p,1,\delta+2}+c_3)\}].$$

We prove (3.11) by induction. It is clear that (3.11) holds for w_0. Assume (3.11) holds for w_{n-1}. Then (3.5), (3.8) and

(3.38) imply that $T_1\|v_{n-1}\|_{1,\infty} \leqq 1$. Hence by (3.6), (3.9), (3.10), (3.37) and (3.38) we can obtain

$$\sup_{t\in[0,T_1]} |w_n(t)|_{p,1,\delta+2} \leq \frac{1}{2}(K_1 + \sup_{t\in[0,T_1]} |w_n(t)|_{p,1,\delta+2}).$$

This proves (3.11). (3.12) follows from (3.5), (3.7) and (3.11).

3.4. Proof of Theorem 1.1.

Let $T_0 = T_1$. Then we have

Lemma 3.7. There exists a unique solution $\{v, w\}$ of (3.1) $\sim$ (3.4) and (2) on $[0,T_0]$, which satisfies

$$\text{(3.39)} \quad \begin{cases} w \in C([0,T_0];M^p_{0,\delta+2}) \cap L^\infty((0,T_0);M^p_{1,\delta+2}), \\ v-v_\infty \in C([0,T_0];M^p_{1,\delta+1}) \cap L^\infty((0,T_0);M^p_{2,\delta+1}). \end{cases}$$

Proof. We shall first show that as $n \longrightarrow \infty$

(3.40) $w_n(x,t)$ converges in $M^p_{0,\delta+2}$ uniformly in $t \in [0,T_0]$,

(3.41) $v_n(x,t)$ converges in $M^p_{1,\delta+1}$ uniformly in $t \in [0,T_0]$.

Let $w^*_n = w_n - w_{n-1}$ and $u_n = v_n - v_{n-1}$. Then by Propositions 3.1 and 3.2 we have

$$\text{(3.42)} \qquad w^*_n \in C([0,T_0];Y^p_{1,\delta+2}), \quad u_n \in C([0,T_0];X^p_{2,\delta+1}).$$

And from (3.5) and (3.6)

$$\text{(3.43)} \qquad \operatorname{rot} u_n = w^*_n,$$

(3.44) $\frac{\partial}{\partial t} w_n^* + \Psi(v_{n-1}, w_n^*) = \Psi(w_{n-1}, u_{n-1});\quad w_n^*(0) = 0,$

where $\Psi(w,u) = (w\cdot\nabla)u - (u\cdot\nabla)w$. It follows from (3.13), (3.42) and (3.43) that for all n

(3.45) $|u_n(t)|_{p,1,\delta+1} \leq c_5|w_n^*(t)|_{p,0,\delta+2}$ for $t\in[0,T_0]$.

Hence from (2.1) and (3.11) we deduce

(3.46) $|\Psi(w_{n-1}(t), u_{n-1}(t))|_{p,0,\delta+2} \leq 2c_1c_5K_1|w_{n-1}^*(t)|_{p,0,\delta+2}$.

Thus (3.44), together with (3.8), (3.9), (3.46) and Lemma 3.3 imply that

(3.47) $|w_n^*(t)|_{p,0,\delta+2} \leq c_6\int_0^t|w_{n-1}^*(s)|_{p,0,\delta+2}ds,$

where $c_6 = 4c_4\{c_1c_5(c_2+1)K_1+c_3\}$. In addition, we have $\|w_1^*\|_{p,0,\delta+2} \leq 2K_1$. Hence repeating (3.47) successively, we obtain

(3.48) $|w_n^*(t)|_{p,0,\delta+2} \leq 2K_1\frac{(c_6t)^{n-1}}{(n-1)!}$ for all n.

From the above estimate we see that w_n satisfies (3.40). Combining (3.45) and (3.48), we have (3.41).

Let $w = \lim_{n\to\infty} w_n$ and $v = \lim_{n\to\infty} v_n$. Then (3.39) follows from (3.40), (3.41) and Lemma 3.3. In addition, taking the limit (3.5) and (3.6), we conclude that $\{v, w\}$ is a solution of the vorticity equation.

To show the uniqueness, suppose that there is another solution $\{\bar{v}, \bar{w}\}$ satisfying the condition of the lemma.

Writing $w^* = w - \overline{w}$ and $u = v - \overline{v}$ and repeating the arguments used to deduce (3.45) and (3.47) from (3.5) and (3.6) we see

$$(3.49)\qquad \begin{aligned} |w^*(t)|_{p,0,\delta+2} &\leq c_6\int_0^t |w^*(s)|_{p,0,\delta+2}ds, \\ |u(t)|_{p,1,\delta+1} &\leq c_5|w^*(t)|_{p,0,\delta+2}. \end{aligned}$$

This proves the uniqueness.

Let $\{v, w\}$ be obtained in the above lemma. Then the velocity v satisfies (2). We shall prove that v satisfies the other conditions. From (3.4), (3.27) and Corollary 3.6 we see that $v(x,0) = v_0(x)$.

Lemma 3.8. $\frac{\partial}{\partial t} v$ exists and belongs to $L^\infty((0,T_0);M^p_{1,\delta+1})$.

Proof. It follows from (3.39) and Lemma 2.3 that $(v\cdot\nabla)v \in L^\infty((0,T_0);M^p_{1,\delta+2})$. Hence Lemma 2.5 implies that there is a solution $q \in L^\infty((0,T_0);M^p_{2,\delta})$ of the Neumann problem:

$$(3.50)\qquad -\Delta q = \operatorname{div}\{(v\cdot\nabla)v\}, \qquad \frac{\partial q}{\partial n}\Big|_S = \{(v\cdot\nabla)v\}\cdot n\big|_S.$$

We put

$$(3.51)\qquad u(t) = v(t) - v_0 + \int_0^t \{(v\cdot\nabla)v + \nabla q - F(\operatorname{rot} f)\}dt.$$

Then it follows from Lemmas 3.5, 3.7 and (3.50) that

$$(3.52)\qquad u \in C([0,T_0];X^p_{1,\delta+1}).$$

Furthermore, we see that rot $u = 0$ holds. Indeed, Lemmas 3.5 and 3.7 imply that for any $\eta \in C_0^\infty(\Omega)$,

$$\frac{d}{dt}(v, \mathrm{rot}\, \eta) = \frac{d}{ds}(w, \eta)$$

$$= (-\{(v\cdot\nabla)v + \nabla q - F(\mathrm{rot}\, f)\}, \mathrm{rot}\, \eta).$$

This, together with $(\mathrm{rot}\, u)|_{t=0} = 0$, proves that rot u = 0. Hence by (3.52) and Corollary 2.7 we have u = 0. This yields

$$\frac{\partial}{\partial t} v = - \{(v\cdot\nabla)v + \nabla q - F(\mathrm{rot}\, f)\}. \tag{3.53}$$

We note that the right-hand side of (3.53) belongs to $L^\infty((0,T_0);M^p_{1,\delta+1})$. This completes the proof.

Let $U = \frac{\partial v}{\partial t} + (v\cdot\nabla)v - f$. Then we have rot U = 0 since v satisfies (3.1), (3.2) and (3.3). In addition, it follows from Lemmas 2.2, 3.7 and 3.8 that $U \in L^\infty((0,T_0);C_b(\bar{\Omega}))$. Hence from Lemma 2.6 there is a scalar function p such that

$$U = - \nabla p \in L^\infty((0,T_0);C_b(\bar{\Omega})). \tag{3.54}$$

This shows that $\{v, p\}$ satisfies (1). The uniqueness follows from Lemma 3.7.

References

[1] Agmon, S., Douglis, A. and Nirenberg, L.: Estimates near the boundary for solutions of elliptic partial differential equations satisfying general boundary conditions. I, Comm. Pure Appl. Math. 12 (1959), 623 - 727.

[2] Bardos, C. and Frisch, U.: Finite-time regularity for bounded and unbounded ideal incompressible fluid using Hölder estimates, Turbulence and Navier Stokes equation, Orsay (1975), Lecture Notes in Mathematics 565, Springer-Verlag, 1 - 14.

[3] Besov, O.V., Il'in, V.P. and Nikol'skii, S.M.: Integral representations of functions and imbedding theorems, vol. II, Winston-Wiley (1979).

[4] Bourguignon, J.P. and Brézis, H.: Remark on the Euler equation, J. Funct. Anal. 15 (1974), 341 - 363.

[5] Cantor, M.: Perfect fluid flows over R^n with asymptotic conditions, J. Funct. Anal. 18 (1975), 73 - 84.

[6] Cantor, M.: Spaces of functions with asymptotic conditions on R^n, Ind. Univ. Math. J. 24 (1975), 897 - 902.

[7] Cantor, M.: Boundary value problems for asymptotically homogeneous elliptic second order operators, J. Diff. Eq. 34 (1979), 102 - 113.

[8] Ebin, D. and Marsden, J.: Groups of diffeomorphisms and the motion of an incompressible fluid, Ann. Math. 92 (1970), 102 - 153.

[9] Günter, N.M.: Potential theory and applications to basic problems of mathematical physics, Frederick Unger Publishing Co. (1967).

[10] Itô, S.: The existence and the uniqueness of regular solution of non-stationary Navier-Stokes equation, J. Fac. Sci., Univ. Tokyo, Sec. I, 9 (1961), 103 - 140.

[11] Judovič, V.: Two-dimensional nonstationary problem of the flow of an ideal incompressible fluid through a given region, Math. Sb. N.S. 64 (1964), 562 - 588 (Russian) = A. M. S. Transl. (2) 57 (1966), 277 - 304.

[12] Kato, T.: On classical solutions of the two-dimensional non-stationary Euler equation, Arch. Rat. Mech. Anal. 25 (1967), 188 - 200.

[13] Kato, T.: Nonstationary flows of viscous and ideal fluids in R^3, J. Funct. Anal. 9 (1972), 296 - 305.

[14] Kato, T. and Lai, C.Y.: Nonlinear evolution equations and the Euler flow, J. Funct. Anal. 56 (1984), 15 - 28.

[15] Kikuchi, K.: Exterior problem for the two-dimensional Euler equation, J. Fac. Sci., Univ. Tokyo, Sec. IA, 30 (1983) 63 - 92.

[16] Kikuchi, K.: The existence and uniqueness of nonstationary ideal incompressible flow in exterior domaions in R^3, to appear.

[17] Nirenberg, L. and Walker, H.: The null spaces of elliptic partial differential operator in R^n, J. Math. Anal. Appl. 42 (1973), 271 - 301.

[18] Swann, H.S.G.: The convergence with vanishing viscosity of nonstationary Navier Stokes flow to ideal flow in R^3, Trans. Amer. Math. Soc. 157 (1971), 373 - 397.

[19] Swann, H.S.G.: The existence and uniqueness of nonstationary ideal incompressible flow in bounded domains in R^3, Trans. Amer. Math. Soc. 179 (1973), 167 - 180.

[20] Temam, R.: On the Euler equations of incompressible perfect fluids, J. Funct. Anal. 20 (1975), 32 - 43.

This Page Intentionally Left Blank

Lecture Notes in Num. Appl. Anal., 8, 73–79 (1985)
Recent Topics in Nonlinear PDE II, Sendai, 1985

Bounds for Solutions of Abstract Nonlinear Evolution Equations

Kyûya Masuda
Mathematical Institute, Tohoku University
Sendai 980, Japan

1. Introduction

The purpose of this paper is to give some bounds for solutions of abstract nonlinear evolution equations. Although we only state an abstract theorem without showing any applications in the present paper, the theorem is very usuful for showing the analyticity of solutions of many different kinds of differential equations such as Korteweg-de Vries equations, Euler equations, nonlinear Schrödinger equations, Navier-Stokes equations, nonlinear diffusion equations in the theory of infiltration. (For the detail see T. Kato-K.Masuda [5].) The abstract evolution equation we consider are of the form

$$\text{(AE)} \qquad d_t u + F(t,u) = 0, \ 0 < t < T \ ; \quad u(0) = u_0$$

($d_t = d/dt$) where F is a nonlinear operator.
To define the F precisely, we first introduce the notion of an S-functional. Let X, Z be two real (or complex) Banach

spaces with $Z \subset X$, the imbedding being continuous. Then we say that a real-valued continuous functional Ψ on $Z \times X$ is an S-functional if the right derivative of $\Phi(v(t))$

$$d_{t+}\Phi(v(t)) = \lim_{h \to 0+} [\ \Phi(v(t+h)) - \Phi(v(t))]/h$$

exists and satisfies the inequality

$$d_{t+}\Phi(v(t)) \leq \Psi(v(t), d_{t+}v(t)), \quad 0 < t < T$$

whenever $v \in C^{+}((0,T);Z) \cap C^{1+}((0,T);X)$, where

$$\Phi(v) = \Psi(v,v)\ , \quad v \in Z.$$

<u>Remark</u>. $C^{+}((0,T);Z)$ is the set of all Z-valued right continous functions on $(0,T)$; and $C^{1+}((0,T);X)$ is the set of all X-valued continuous functions on $(0,T)$, having the right derivative in $(0,T)$.

We show two examples of S- functionals.

<u>Examples</u>

1. In connection with the existence problem for a system of ordinary differential equations, Kamke [3] introduced the notion of an S-function. (see also Hukuhara [2]). A real-valued function $S(x)$ on R^n is called the <u>S-function</u> if

a) $S(x)$ is continuous on R^n ;

b) $S(0) = 0$;

c) the inequality

$$d_{t\pm}\, S(v(t)) \leq S(d_{t\pm} v(t))$$

holds for all left (right) differentiable functions $v(t)$ with values in R^n. (d_{t-} denotes the left derivative.) Then the Ψ defined by $\Psi(v,w) = S(w)$ $(v, w \in R^n)$ is an S-functional on $R^n \times R^n$.

2. Let $\{X,Y,Z\}$ be an admissible triplet(of Banach spaces). (For the definition see Kato-Lai [4].) Let $\langle \cdot , \cdot \rangle_{Z\times X}$ be a non-negative continuous bilinear form associated with the above triplet. Then the $\langle \cdot , \cdot \rangle_{Z\times X}$ is an S-functional on $Z \times X$.

We next define an S-family. A family $\{\Psi_\sigma\}$, $0 < \sigma < \sigma^*$, of S-functionals Ψ_σ on $Z \times X$ will be called the S-family if

$\Phi_\sigma(v)$ ($= \Psi_\sigma(v,v)$) is continuous in σ ($0 < \sigma < \sigma^*$) and v ($\in Z$) and if for each $v \in Z$ $\Phi_\sigma(v)$ is differentiable in σ, and its derivative $\partial_\sigma \Phi_\sigma(v)$ is continuous in σ and v ($\in Z$).

Now consider a somewhat more general equation than (AE):

$$\text{(AE')} \qquad d_{t+}\, u + F(t,u) = 0, \quad 0 < t < T, \quad u(0) = u_0.$$

With above notations our theorem now reads:

THEOREM Let $\{\Psi_\sigma\}$, $0 < \sigma < \sigma^*$, be an S-family. Let $F = F(t,v)$ be a mapping of $[0,T) \times Z \to X$ satisfying

(1) $$\Psi_\sigma(v,-F(t,v)) \le \sigma\,\alpha(\Phi_\sigma(v))\partial_\sigma\Phi_\sigma(v) + \beta(\Phi_\sigma(v))$$

($\Phi_\sigma(v) = \Psi_\sigma(v,v)$) for $v \in Z$, $0 \le t < T$, and $0 < \sigma < \sigma^*$,

where $\alpha(s)$, $\beta(s)$ are given continuous functions of s.

Let $u_0 \in Z$. Suppose that $u = u(t)$ is a solution of (AE') with

(2) $$u \in C^+([0,T);Z) \cap C^{1+}((0,T);X).$$

Then for any σ_0 ($0 < \sigma < \sigma^*$) there is a T_0 ($0 < T_0 < T$) such that

(3) $$\Phi_{\sigma(t)}(u(t)) \le z(t), \quad 0 \le t < T_0$$

where $z = z(t)$ is the maximal solution of the ordinary differential equation

(4) $$d_t z = \beta(z) \;; \qquad z(0) = \Phi_{\sigma_0}(u_0)$$

and

(5) $$\sigma(t) = \sigma_0 \exp[-\int_0^t \alpha(z(s))ds].$$

Remark The $\sigma = \sigma(t)$ is defined on the maximal interval $[0, T_{max})$ of existence of solution z. The T_0 is such that the interval $[0,T_0)$ is the maximal interval on which $\sigma(t) < \sigma^*$.

2. Proof of Theorem

The proof is simple, done by the method of characteristics. We first consider the first order partial differential inequality of the form

$$\partial_{t+} y \leq \sigma a(t,y)\partial_\sigma y + b(t,y), \quad 0<\sigma<\sigma^*, \quad 0<t<T, \tag{6}$$

where a, b are given continuous functions on $[0,T) \times R^1$, and $\partial_{t+}y$ denotes the right partial derivative of y with respect to t :

$$\partial_{t+} y(\sigma,t) = \lim_{h\to 0+} [y(\sigma,t+h)-y(\sigma,t)]/h.$$

Then we want to estimate $y(\cdot,t)$ in terms of $y(\cdot,0)$.

LEMMA Let $y = y(\sigma,t)$ be a solution of (6). Then for every σ_0

$$y(\sigma(t),t) \leq z(t), \quad 0 \leq t < T_0 \tag{7}$$

where $z = z(t)$ is the maximal solution of the ordinary differential equation

$$d_t z = b(t,z), \quad z(0) = y(\sigma_0,0); \tag{8}$$

$$\sigma(t) = \sigma_0 \exp[-\int_0^t a(s,z(s))ds] ; \tag{9}$$

and T_0 is the supremum of all T' such that $\sigma(t) < \sigma^*$

on $[0, T')$.

Remark By a solution y of (6) we mean a continuous function $y = y(\sigma,t)$ of σ, t ($0 < \sigma < \sigma^*$, $0 \leq t < T$) with

i) the right derivative $\partial_{t+}y$ and usual partial derivative $\partial_\sigma y$ in σ both exist and are continuous in σ, t.

ii) $y = y(\sigma,t)$ satisfies the inequality (6).

Proof of Lemma. Given $\varepsilon > 0$, let $z_\varepsilon = z_\varepsilon(t)$ be the maximal solution of the equation

$$d_t z = b(t,z) + \varepsilon \text{ , } z(0) = y(\sigma_0\ 0) + \varepsilon.$$

Let $\sigma_\varepsilon = \sigma_\varepsilon(t)$ be the function defined by (9) with z replaced by the z_ε defined above. Then the function

$$w(t) = y(\sigma_\varepsilon(t),t) - z_\varepsilon(t)$$

satisfies the inequality

$$d_{t+}w \leq f(t,w) - \varepsilon, \quad w(0) = -\varepsilon$$

where

$$f(t,w) = b(t,w+z_\varepsilon) - b(t,z_\varepsilon) +$$

$$+ \sigma_\varepsilon(t)[a(t,w+z_\varepsilon)-a(t,z_\varepsilon)]\ \partial_\sigma y(\sigma_\varepsilon(t)\ t).$$

Hence it easily follows from the comparison theorem for a differential inequality that $w(t) < 0$, and so that

(10) $$y(\sigma_\varepsilon(t),t) < z_\varepsilon(t), \quad 0 < t < T_\varepsilon.$$

(T_ε is as T in Lemma.) On the other hand it follows from Kamke's theorem (see [1]) that there is a subsequence$\{\varepsilon_j\}_{j=1}^{\infty}$ tending to zero, through which $\{z_\varepsilon(t)\}$ converges to a solution z_0 of (8) uniformly on any compact subinterval of $[0,T_0)$. Using the comparison theorem, one can conclude that the z_0 is the maximal solution. Thus by (10) we have (7). This completes the proof of Lemma.

Poor of Theorem It is now straightforward. On setting

$$y(\sigma,t) = \Phi_\sigma(u(t)),\ a(t,y) = \alpha(y),\ b(t\ y) = \beta(y)$$

in the Lemma, the Theorem immediately follows from the Lemma.

REFERENCES

[1] P.Hartman, Ordinary Differential Equations, John Wiley and Sons, 1964.

[2] M. Hukuhara, Sur la fonction S(x) de M.E,Kamke, Japanese Journal of Mathematics, 17 (1941), 289-298.

[3] E.Kamke, Über die eindeutig Bestimmtheit der Integrale von Differentialgleichungen, Sitzunsber.Akad.Heidelberg, 11 (1931), 10-15.

[4] T.Kato and C.Y.Lai, Nonlinear evolution equations and the Euler flow, J.Funct. Anal., 56 (1984), 15-28.

[5] T.Kato and K.Masuda, Nonlinear evolution equations and Analyticity. (pre·print).

This Page Intentionally Left Blank

Lecture Notes in Num. Appl. Anal., 8, 81–105 (1985)
Recent Topics in Nonlinear PDE II, Sendai, 1985

On Prandtl Boundary Layer Problem

Shin'ya MATSUI and Taira SHIROTA
Department of Mathematics, Hokkaido University
Sapporo, 060, Japan

1. Introduction.

In this paper we study two mathematical bases of the boundary layer problem for two dimensional incompressible laminar viscous flow. The one of them is the existence of a separation point and the other is related to the approximation to the Navier-Stokes flow before such separation points. The equations of this problem along a rigid wall are

$$(1.1)\qquad \begin{aligned} uu_x + vu_y &= \nu u_{yy} - p_x, \\ u_x + v_y &= 0, \end{aligned}$$

in the domain $D_A = \{(x,y);\ 0 < x < A,\ 0 < y < \infty\}$. Here (x,y) are orthogonal coordinates in the boundary layer with x representing the length along the wall and y the perpendicular distance from the wall (The subscripts in (1.1) denote the partial differentiation with respect to the corresponding variable). $(u(x,y), v(x,y))$ is the unknown velocity vector field, $p_x = p_x(x)$ is a prescribed pressure gradient and the constant ν is the kinematic viscosity.

For a given exterior flow $U = U(x)$, the appropriate boundary conditions are

$$(1.2)\qquad \begin{aligned} &u = v = 0 \quad \text{for} \quad y = 0, \\ &u(x,y) \to U(x) \quad \text{as} \quad y \to \infty \quad \text{uniformly in } x \text{ on any} \\ &\text{compact subset of } [0,A). \end{aligned}$$

In order to obtain well-set problem, we suppose that at an initial position, say $x = 0$, a datum $u_0(y)$ is assigned the velocity component u, i.e.,

$$(1.3)\qquad u(0,y) = u_0(y) \quad \text{for} \quad y \in [0,\infty).$$

Here we assume the Bernoulli law and, for convenience of discussion, that the initial position is not a stagnation one, i.e.,

$$(1.4)\qquad \begin{aligned} &U(x)U_x(x) + p_x(x) = 0, \\ &U(0) > 0. \end{aligned}$$

In Section 2 we shall present how to obtain the solution of the boundary layer problem until its separation point, whenever a stagnation point occurs downstream. Furthermore we shall also prove that the separation points appear before the stagnation point ([8]). In physical and enginering points of view these results are well known as experimental facts ([7]). In fact, in the Goldstein's research he assume the results above. (For the regularity assumption mentioned below in (2.1), (2.2) and the Goldstein's one see ([6], [16])).

In section 3, as the applications of the above, first we shall show that if the pressure gradient is positive at some downstream point, then there exists a solution with its separation point. This result contrasts with the Serrin's ([15]) and Nickel's ([10]). In Section 3.2 we shall try to obtain some rigorous connection between the Navier-Stokes equations and the Prandtl equations along the Fife's considerations ([1]) (in the case where the pressure gradient $p_x^\nu \le -a < 0$ for some positive constant a). In our case the gradient p_x^ν is not always negative. But we must assume that the Navier-Stokes flow has no separation point in a certain strong sense (see (v) in Φ below).

Though all of our methods of considerations are simple, the results will be important as the mathematical bases of the boundary layer theory (for instance, see [2]).

2. The existence of separation points.

2.1. Preliminaries.

In this section we assume the following

(2.1) The exterior flow $U(x)$ is sufficiently smooth, vanishes at some point $x = X_0$ $(0 < X_0 < \infty)$ and is positive for $x \in [0,X_0)$.

(2.2) If $p_x(c) = 0$ at a point $x = c$, then a certain N-th derivative of p_x does not vanish at this point.

We define the spaces $I^{2+\alpha}$, $P^2([0,A))$ used in this paper and the separation point. For an interval $[a,A]$ let $B^0([a,A] \times [0,\infty))$ be the Banach space of uniformly continuous functions in $[a,A] \times [0,\infty)$ with supremum norm. For any $\alpha(0 < \alpha \le 2/3)$ and $y_0 > 0$ let $C^\alpha([a,A] \times [y_0,\infty))$ be the Banach space of Hölder continuous functions, i.e.,

$$|u(x_1,y_1) - u(x_2,y_2)| \le M\{|x_1-x_2|^{1/2} + |y_1-y_2|\}^\alpha$$

for $(x_i,y_i) \in [a,A] \times [y_0,\infty)$ $(i=1,2)$ and $M = M(y_0,u)$, and let $C^\alpha([a,A] \times (0,\infty)) = \bigcap_{y_0>0} C^\alpha([a,A] \times [y_0,\infty))$. Furthermore we define $B^\alpha([a,A] \times (0,\infty)) =$

$B^0([a,A] \times [0,\infty)) \cap C^\alpha([a,A] \times (0,\infty))$. We also define $B^0([0,\infty))$, $C^\alpha((0,\infty))$ and $B^{2+\alpha}((0,\infty))$ by the analogous way.

Then we define the space of the initial data :

$$\begin{aligned} I^{2+\alpha} &= I^{2+\alpha}(\nu,U) \\ &= \{u(y) \in B^{2+\alpha}((0,\infty)) \ ; \\ &\quad u(0) = 0,\ u_y(0) > 0,\ u_y(y) \geq 0 \text{ for } y \in [0,\infty), \\ &\quad u \to U(0) \text{ as } y \to \infty \text{ and} \\ &\quad \nu u_{yy}(y) - p_x(0) = O(y^2) \text{ as } y \to 0\}. \end{aligned}$$

The condition in $I^{2+\alpha}$

$$\nu u_{yy}(y) - p_x(0) = O(y^2) \quad \text{as} \quad y \to 0 \tag{2.3}$$

is a (strong) compatibility condition.

The space of the solutions is given as follows :

$(u,v) \in P^2([a,A])$ means that

(i) $u,\ u_y, u_{yy} \in B^0([a,A] \times [0,\infty))$,

(ii) $u_x,\ v,\ v_y \in C^0([a,A] \times [0,\infty))$,

(iii) $u(x,y) > 0$ in $[a,A] \times (0,\infty)$,

(iv) $u_y(x,0) > 0$ for $x \in [a,A]$.

Then we define the space by

$$P^2([a,A)) = \bigcap_{0<A'<A} P^2([a,A']).$$

Now we define *the separation point* of a solution to the problem (1.1), (1.2) and (1.3) :

Definition. A point $(s,0)$ is a separation point of a solution (u,v) to our problem in the domain D_s, if $(u,v) \in P^2([0,s))$ and for some sequence (x_n,y_n) in $[0,s) \times [0,\infty)$

$$(x_n,y_n) \to (s,0) \quad \text{and} \quad u_y(x_n,y_n) \to 0 \quad \text{as} \quad n \to \infty.$$

2.2. Results and Remarks.

Theorem 1. ([8]) For the problem (1.1), (1.2) and (1.3) with the initial data $u_0(y) \in I^{2+\alpha}$ there exists a solution $(u,v) \in P^2([0,s))$ such that $(s,0)$ is its separation point. Furthermore if we put $S(u_0) = s$,

$$\sup\{S(u_0);\ u_0 \in I^{2+\alpha}\} < X_0 \tag{2.4}$$

holds.

Here we note that physicists call $(s,0)$ a separation point if $u_y(s,0) = 0$ for the solution (u,v) (asymptotically) (see [7]). On the other hand the separation point $(S(u_0),0)$ in Theorem 1 is the same as the above, if u and u_y belong to $C^0([0,s] \times [0,\infty))$. But to our knowledge, the existence of such an exact solution is not proved under our assumptions yet. Furthermore the separation point called by physicists does not depend on the viscosity. Therefore in a certain case the point $x = \sup\{S(u_0): u_0 \in I^{2+\alpha}(1,U)\}$ may be thought as the separation point physicists call.

For convenience of studying we consider the transformation of the independent variables in the system (1.1) of the form

$$x = x, \quad \psi = \psi(x,y), \tag{2.5}$$

where

$$\psi_y(x,y) = u, \quad \psi_x(x,y) = -v, \quad \psi(x,0) = 0.$$

That is, the stream lines of (1.1) are thought as the new coordinates. If we put $\omega(x,\psi) = u^2(x,y)$, then (2.5) reduces the problem (1.1), (1.2) and (1.3) to the Von Mises' form :

$$L(\omega) = \nu\sqrt{\omega}\,\omega_{\psi\psi} - \omega_x = 2p_x \quad \text{in} \quad G_A \tag{2.6}$$

with the conditions :

$$\begin{aligned} &\omega(x,0) = 0, \\ &\omega(0,\psi) \to U^2(x) \quad \text{as} \quad \psi \to \infty \quad \text{uniformly in} \quad x \in [0,A), \end{aligned} \tag{2.7}$$

$$\omega(0,\psi) = \omega_0(\psi), \tag{2.8}$$

where $G_A = \{(x,\psi);\ 0 < x < A,\ 0 < \psi < \infty\}$ and $\omega_0(\int_0^y u_0(t)dt) = u_0^2(y)$. Hereafter we always assume that the initial data $\omega_0 \in I_M^{2+\alpha}$ for some α, where

$$I_M^{2+\alpha} = \{\omega(\psi);\ \omega(\int_0^y u(t)dt) = u^2(y),\ u \in I^{2+\alpha}\}.$$

Then $\omega(\psi) \in I_M^{2+\alpha}$ if and only if ω, ω_ψ and $\sqrt{\omega}\,\omega_{\psi\psi} \in B^\alpha((0,\infty))$, $\omega(0) = 0$, $\omega_\psi(0) > 0$, $\omega_\psi(\psi) \geq 0$ for $\psi \geq 0$, $\omega(\psi) \to U^2(0)$ as $\psi \to \infty$ and ω satisfies the compatibility condition :

$$\mu(\psi) = \nu\sqrt{\omega}\,\omega_{\psi\psi}(\psi) - 2p_x(0) = O(\psi) \quad \text{as} \quad \psi \to 0. \tag{2.9}$$

Let $P_M^{2+\alpha}([0,A]) = \{\omega(x,\psi):\ \omega,\ \omega_x,\ \omega_\psi,\ \sqrt{\omega}\,\omega_{\psi\psi} \in B^\alpha([0,A] \times (0,\infty))$,

$|\omega_x| \leq K\psi^{1-\beta}$ and $\omega_\psi \geq m$ in $[0,A] \times [0,\psi_1]$ and $\omega(x,\psi) \geq k$

for $\psi \geq \psi_1\}$,

where positive constants ψ_1, m, k depend on ω. Furthermore for any β ($0 < \beta < 1/2$) the positive constant K depends on ψ_1, β and ω. Finally we define $P_M^{2+\alpha}([0,A)) = \bigcap_{0<A'<A} P_M^{2+\alpha}([0,A'])$.

Here we summarize the results and remarks on the local solutions of

(1.1), (1.2), (1.3) obtained by Oleinik ([11] and [14]).

(a) For some $A_0 > 0$ the local solution $\omega(x,\psi) \in P_M^{2+\alpha}([0,A_0])$ of (2.6), (2.7) and (2.8) in G_{A_0} is given by the following way (c.f. Lemma 4). For small $\varepsilon > 0$, let $G_A^\varepsilon = \{(x,\psi);\ 0 < x < A,\ 0 < \psi < 1/\varepsilon\}$. Then there exists the approximate positive solution $\omega_\varepsilon(x,\psi)$ in $G_{A_0}^\varepsilon$, which satisfies (2.6) with boundary conditions :

$$\begin{aligned}
&\omega_\varepsilon(0,\psi) = \omega_0(\varepsilon+\psi),\\
(2.10)\qquad &\omega_\varepsilon(x,0) = \omega_0(\varepsilon)\exp\{\mu(\varepsilon)x/\omega_0(\varepsilon)\},\\
&\omega_\varepsilon(x,1/\varepsilon) = \omega_0(\varepsilon+1/\varepsilon)\exp\{\mu(\varepsilon+1/\varepsilon)x/\omega_0(\varepsilon+1/\varepsilon)\}.
\end{aligned}$$

Furthermore the solution $\omega_\varepsilon(x,\psi)$ belongs to $P_M^{2+\alpha}(\overline{G}_{A_0}^\varepsilon)$ for $\alpha \leq 2/3$, where the last space is defined as in $P_M^{2+\alpha}([0,A_0])$, but we must replace $B^\alpha([0,A] \times (0,\infty))$ by $B^\alpha([0,A_0] \times (0,1/\varepsilon])$. And the set $\{\omega_\varepsilon(x,\psi)\}_{\varepsilon \ll 1}$ has the following properties :

(2.11) All constants in $P_M^{2+\alpha}$ (i.e., α, m, ψ_1, K and k) and supremum norms over $\overline{G}_{A_0}^\varepsilon$ of the functions ω_ε, ω_{ε_x}, $\omega_{\varepsilon_\psi}$, $\sqrt{\omega_\varepsilon}\,\omega_{\varepsilon_{\psi\psi}}$ and $\omega_\varepsilon^{\beta-1}\omega_{\varepsilon_x}$ do not depend on ε ($\ll 1$).

Then there exists a subsequence $\{\omega_{\varepsilon'}\}$ such that $\omega_{\varepsilon'} \to \omega$ in $C^0([0,A_0] \times [0,N])$ for any $N \gg 1$ and $\omega_{\varepsilon'_x}$, $\omega_{\varepsilon'_\psi}$, $\sqrt{\omega_{\varepsilon'}}\,\omega_{\varepsilon'_{\psi\psi}}$ converge to the corresponding functions of the unique local solution ω in $C^0([0,A_0] \times [1/N,N])$ as $\varepsilon' \to 0$. Here we note that $\omega(x,\psi) \to U^2(x)$ as $\psi \to \infty$ uniformly in x on $[0,A_0]$.

(b) Oleinik assumes that the initial data u_0 belong to $B^{2+\alpha}([0,\infty))$, but ours belong to $B^{2+\alpha}((0,\infty))$. Our assumption of regularity is enough for the existence of the local solution.

(c) A solution $(u,v) \in P^2([0,A])$ of (1.1), (1.2) with initial data

$u_0 \in I^{2+\alpha}$ exists, if a solution $\omega \in P_M^{2+\alpha}([0,A])$ of (2.6), (2.7) and (2.8) with the corresponding initial data ω_0 by (2.5) exists.

Now by the transformation (2.5), we have $\omega_\psi = 2u_y$. Therefore we define *the separation point* of the solution ω to the problem (2.6), (2.7) and (2.8) by the same way as the above, i.e., a point (s,0) is the separation one of a solution $\omega(x,\psi)$, if $\omega \in P_M^{2+\alpha}([0,s))$ and there exists a sequence $(x_n,\psi_n) \in [0,s) \times [0,\infty)$ such that

$$(x_n,\psi_n) \to (s,0) \quad \text{and} \quad \omega_\psi(x_n,\psi_n) \to 0 \quad \text{as} \quad n \to \infty.$$

We also denote this separation point by $S(\omega_0)$. Then, in fact, we obtain that $S(u_0) = S(\omega_0)$ for u_0 corresponding to ω_0 by (2.5).

2.3. Lemmas and Proof of Theorem 1.

Lemma 1. Let $\omega(x,\psi)$ be the local solution mentioned in Section 2.2-(a). Then there exist positive constants M and λ such that

$$|\omega_x| \leq M\psi \quad \text{for} \quad 0 \leq x \leq A_0 \quad \text{and} \quad 0 \leq \psi \leq \lambda. \tag{2.12}$$

Proof. Let $\omega_\varepsilon(x,\psi)$ be the approximate solution given in Section 2.2-(a). We put $E = \{(x,\psi): 0 \leq x \leq A_0,\ 0 \leq \psi \leq \lambda\}$, where the samll positive constant λ will be determined later. From the fact (a), if we show

$$|\omega_\varepsilon^{-1}\ \omega_{\varepsilon_x}| \leq M \quad \text{in} \quad E$$

for some M, λ and ε (<< 1), then we obtain (2.12). Hence we will show the above inequality.

Since $\omega_\varepsilon > 0$ (see (a)), we may put $g_\varepsilon(x,\psi) = \ln(\omega_\varepsilon(x,\psi))$, then from (2.6) g_ε satisfies the following equation (we omit subscript ε in functions for convenience of discriptions).

(2.13) $$\nu g_{\psi\psi}e^{3g/2} + \nu(g_\psi)^2 e^{3g/2} - e^g g_x = 2p_x.$$

Differentiating (2.13) with respect to x, we obtain an equation for $h = g_x$:

$$\nu h_{\psi\psi}e^{3g/2} + \frac{3}{2}\nu e^{3g/2}g_{\psi\psi}h + \frac{3}{2}\nu e^{3g/2}(g_\psi)^2 h +$$
$$+ 2\nu e^{3g/2}g_\psi h_\psi - e^g h^2 - e^g h_x = 2p_{xx}.$$

Replaceing $\nu g_{\psi\psi}$ exp 3g/2 by its value from (2.13), the above equation for h is written in the form

(2.14) $$\nu e^{3g/2}h_{\psi\psi} + Ch + 2\nu e^{3g/2}g_\psi h_\psi - e^g h_x = 2p_{xx},$$

where $C = \frac{1}{2}he^g + 3p_x = \frac{1}{2}\omega_x + 3p_x$. From (2.11),

(2.15) $$|C| \leq M_1 \quad \text{in} \quad \overline{G}^\varepsilon_{A_0}$$

for some positive constant M_1 (independent of ε). Therefore (2.14) implies that $j = h\cdot\exp\{-(M_1+1)x\}$ satisfies

(2.16) $$\nu e^{3g/2}j_{\psi\psi} + (C-M_1-1)j + 2\nu e^{3g/2}g_\psi j_\psi - e^g j_x = 2p_{xx}e^{-(M_1+1)x}.$$

Now if j attains its maximum in E without the lines $x = 0$, $\psi = 0$ and $\psi = \lambda$, then at this point $\nu e^{3g/2}j_{\psi\psi} - e^g j_x \leq 0$ and $j = 0$ hold. Hence from (2.15) and (2.16) we deduce

(2.17) $$j \leq \frac{2p_{xx}\cdot\exp\{-(M_1+1)x\}}{C - M_1 - 1} \leq M_2$$

at the maximum point, where the positive constant M_2 does not depend on ε. By the same arguments, at the minimum point of j in the same subdomain of E

we have

(2.18) $$j \geq -M_2.$$

If ε and λ are sufficiently small, from (2.9) it follows that $h(0,\psi) = \omega_0(\varepsilon+\psi)^{-1}\mu(\varepsilon+\psi) = O(1)$ and $h(x,0) = \omega_0(\varepsilon)^{-1}\mu(\varepsilon) = O(1)$. Furthermore (2.11) implies $h(x,\lambda) = O(1)$.

Thus from (2.17) and (2.18) we conclude that

$$|h| \leq M_3 \quad \text{in} \quad E,$$

where the positive constant M_3 does not depend on ε. This proves Lemma 1.

q.e.d.

Using the maximum principle (for instance, see Theorem 8 of the section 5 in [5]), we have

<u>Lemma 2</u>. ([10] or [8]) Let $\omega(x,\psi)$ be a solution of (2.6), (2.7) and (2.8) belonging to $P_M^{2+\alpha}([0,A])$. Then $\omega(x,\psi)$ is monotone nondecreasing with respect to ψ in $\overline{G}_A$.

It is easy to see that Lemma 1 and 2 imply the following

<u>Corollary 1</u>. Let $\omega(x,\psi)$ be the local solution mentioned in Section 2.2-(a). Then the section $\omega(x,\cdot)$ belongs to $I_M^{2+\alpha}$ for any x $(0 \leq x \leq A_0)$.

<u>Lemma 3</u>. There exists no solution $\omega(x,\psi)$ of (2.6), (2.7) and (2.8), which belongs to $P_M^{2+\alpha}([0,X_0))$ and whose section $\omega(x,\cdot)$ belongs to $I_M^{2+\alpha}$ $(0 \leq x < X_0)$.

Proof. First we note that from our assumptions (2.1) (2.2) and the Bernoulli law (1.4) there exists a constant d (> 0) such that

(2.19) the pressure gradient $p_x(x)$ is monotone non increasing in $[X_0-d,X_0]$.

Next for a certain $A \in (X_0-d, X_0)$ determined later, let $\omega \in P_M^{2+\alpha}([0,A])$ such that

$$\omega(X_0-d,\psi) \leq d^{-1/2} m\psi \int_{X_0-d}^{X_0} p_x(t)dt \quad \text{for} \quad 0 \leq \psi \leq 2\sqrt{d}/m,$$

then we see below that

$$\omega(x,\psi) \leq F(x,\psi) \equiv \tag{2.20}$$

$$\frac{2m}{\sqrt{d}}\psi\left\{\int_{X_0-\frac{d(A-x)}{A-X_0+d}}^{X_0} p_x(t)dt\left(1-\frac{\gamma\psi}{\sqrt{d}}\right)+\frac{\gamma\psi}{\sqrt{d}}\int_x^{X_0} p_x(t)dt\right\}$$

on $[X_0-d,A] \times [0,\sqrt{d}/\gamma]$, for some positive constants m and γ. Using (2.20) and $\omega(x,0) = 0$, we obtain

$$\frac{\omega(x,\psi)-\omega(x,0)}{\psi} \leq \frac{2m}{\sqrt{d}}\int_{X_0-\frac{d(A-x)}{A-X_0+d}}^{X_0} p_x(t)dt\left(1-\frac{\gamma\psi}{\sqrt{d}}\right)$$

$$+\frac{2\gamma\psi}{d}\int_x^{X_0} p_x(t)dt$$

for $X_0 - d \leq x \leq A$ and $0 \leq \psi << 1$. Therefore if $\psi \downarrow 0$ and $x \uparrow A$, then we have $\omega_\psi(A,0) \leq 0$. This contradicts our assumption $\omega_\psi(x,0) > 0$ for $0 \leq x < X_0$ (see the definition of $P_M^{2+\alpha}$).

Now we show (2.20) in the case where $X_0 = d = 1$.

Since $\omega(0,\psi) \in I_M^{2+\alpha}$, we may find a positive constant m such that

$$\omega(0,\psi) \leq m\psi\int_0^1 p_x(t)dt \quad \text{for} \quad 0 \leq \psi \leq 2/m. \tag{2.21}$$

Furthermore (2.21) implies that

(2.22) $$\omega(0,\psi) \le 2(m\psi - m^2\psi^2/4)\int_0^1 p_x(t)dt \quad \text{for} \quad 0 \le \psi \le 2/m.$$

Here we define

$$H(x,\psi) = 2(m\psi - m^2\psi^2/4)\int_x^1 p_x(t)dt \quad \text{for} \quad 0 \le \psi \le 2/m.$$

Then since $\omega(x,\psi) \le U^2(x) = 2\int_x^1 p_x(t)dt$ (using monotonicity of ω and (1.4)) it follows from (2.21) that an inequality $\omega(x,\psi) \le H(x,\psi)$ holds on the lines $x = 0$, $\psi = 0$ and $\psi = 2/m$. Furthermore by simple calculation we obtain $L(H) - L(\omega) \le 0$ in $[0,A] \times [0,2/m]$ for any $A < 1$. Hence by the maximum principle we have

(2.23) $$\omega(x,\psi) \le H(x,\psi) \quad \text{on} \quad [0,A] \times [0,2/m]$$

(see the proof of Theorem 3 in [11]).

Now from (2.21) and (2.23), the inequality (2.20) holds on the lines $x = 0$, $\psi = 0$ and $\psi = 1/\gamma$ for $2\gamma > m$. Furthermore, from the monotonicity of p_x and (2.1), we deduce that

$$\begin{aligned} L(F) &= \nu\sqrt{F}\, F_{\psi\psi} - F_x \\ &\le 4\nu m\gamma p_x(x)\cdot(x/A - x)\left(2m\gamma^{-1}\int_0^1 p_x(t)dt\right)^{1/2} + 2m\psi p_x(x)/A \\ &\le 2cm^{3/2}\gamma^{1/2}(1-A)p_x(x) + 2mp_x(x)/\gamma A, \end{aligned}$$

for $0 \le \psi \le 1/\gamma$, $0 \le x \le A$ and any γ ($> m/2$), A (< 1), where $c^2 = 8\nu^2\int_0^1 p_x(t)dt$.

Hence an inequality $L(F) \le 2p_x(x)$ holds for $0 \le x \le A$ and $0 \le \psi \le 1/\gamma$ if an inequality

(2.24) $$cm^{3/2}\gamma^{1/2}(1-A)A + m/\gamma \leq A$$

holds.

Therefore considering the quadraic inequality in A, we may find such A (< 1) for $\gamma > m$. Thus by the maximum principle we have the inequality (2.20) in this case.

Finally by the coordinate transformation

(2.25) $$\tilde{x} = (x-X_0+d)/d, \quad \tilde{\psi} = \psi/\sqrt{d}\ ,$$

the general case reduces to the previous one. Thus we obtain Lemma 3.

q.e.d.

From the above argument, especially (2.24) and (2.25), we have the following

Corollary 2. For the separation point (s,0) we have the following a priori estimate :

(2.26) $$0 < s \leq A = X_0 - d(1-\tilde{A}),$$

where $\tilde{A} = 1 - 2/(1+D^{1/2})$, $D = 1 + 3^3c^2m^4$ and $c^2 = 8\nu^2\int_{X_0-d}^{X_0} p_x(t)dt$.

For a point $x = X_1 \in (0,X_0)$ let k_1 and k_2 be $\min_{0\leq x\leq X_1} U^2(x)$ and $\max_{0\leq x\leq X_1} |2p_x(x)|$ respectively. Furthermore for sufficiently small positive constants ψ_0 and k, let W be the subset of $I_M^{2+\alpha}$ such that $\omega_0(\psi) \in W$ satisfies

$$\inf\{\omega_{0_\psi}(\psi);\ 0 \leq \psi \leq \psi_0\} \geq k.$$

Then we obtain the following lemma which is essentially given by

Oleinik [11].

Lemma 4. For any $\omega_0 \in W$, the constant A_0 in Section 2.2-(a) may be chosen dependently only on ψ_0, k, k_1 and k_2, but independently of ω_0.

Proof of Theorem 1. Let $\omega(x,\psi)$ be the local solution in G_{A_0}. From Corollary 1 and the facts (a) and (b) in Section 2.2 there exists a solution which is a continuation of the local solution and belongs to $P_M^{2+\alpha}([0,A])$ for some $A > A_0$ (By $\omega(x,\psi)$ we also denote this continuation). Furthermore from Corollary 1 we see also that its section $\omega(x,\cdot)$ belongs to $I_M^{2+\alpha}$ for any x $(0 \leq x \leq A)$.

Let $s = \sup\{A$; the above continuation exists in $G_A\}$. Then from Lemma 3 we see that $s < X_0$. Assume that $(s,0)$ is not the separation point. That is, for sequence $A_n < s$ with $A_n \to s$ (as $n \to \infty$) there exist positive constants m_0, ψ_0 and a natural number n_0 which do not depend on n and satisfy

$$(2.27) \qquad \inf\{\omega_\psi(A_n,\psi);\ 0 \leq \psi \leq \psi_0\} \geq m_0 \quad \text{for} \quad n \geq n_0.$$

For an arbitrary but fixed $n \geq n_0$, if we consider the point $x = A_n$ as an initial position and the section $\omega(x,\cdot)$ as an initial datum, there exists the solution of (2.6), (2.7) and (2.8) in G_{A_n+B}, which belongs to $P_M^{2+\alpha}([A_n, A_n+B])$ for some $B > 0$. On the other hand from Lemma 4 and (2.27) the constant B does not depend on $n \geq n_0$. Hence, for sufficiently large n_1 an inequality $A_{n_1}+B > s$ holds. Thus we get a solution of (2.6), (2.7) and (2.8) in $G_{A_{n_1}+B}$ as a continuation of the solution $\omega(x,\psi) \in P_M^{2+\alpha}([0,s))$, which is contrary to the definition of s. This proves the first part of Theorem 1.

The second assertion of Theorem 1, i.e. (2.4), is proved by the following way. First we mention the relation between the Blasius solution and our solution obtained above. The Blasius solution is given as follows (see [3] and [4]).

Let $f \equiv f(\eta)$ be the solution of

$$f''' + ff'' = 0 \quad \text{in} \quad [0,\infty),$$

with conditions $f(0) = f'(0) = 0$ and $f'(\eta) \to 1$ as $\eta \to \infty$.

We put in (2.5)

$$\psi(x,y) = \{2\nu(U(X_0-d) + e)\}^{1/2} x^{1/2} f(\eta),$$

where $\eta = \left(\dfrac{U(X_0-d)+e}{2\nu x}\right)^{1/2} y$, the constant d is the same one in the proof of Lemma 3 and e is a positive constant.

Then

$$\omega_B(x,\psi) = \{U(X_0-d) + e\}^2 (f'(\eta))^2$$

satisfies

$$L(\omega_B) = 0 \quad \text{in} \quad (0,\infty) \times (0,\infty),$$

$$\omega_B(x,0) = 0 \quad \text{for} \quad x > 0,$$

$$\omega_B(x,\psi) \to \{U(X_0-d) + e\}^2 \quad \text{as} \quad \psi \to \infty, \text{ pointwise} \quad \text{in} \quad x > 0,$$

$$\omega_B(x,\psi) \to \{U(X_0-d) + e\}^2 \quad \text{as} \quad x \to 0, \text{ pointwise} \quad \text{in} \quad \psi > 0.$$

Now let $\omega(x,\psi)$ be the solution obtained above with $S(\omega_0) > X_0 - d$.

Then we have

$$(2.28) \qquad \omega(x,\psi) \le \omega_B(x-X_0+d,\psi) \quad \text{in} \quad [X_0-d, S(\omega_0)] \times [0,\infty).$$

We note that from Corollary 2 $S(\omega_0) < X_0$ holds. The proof of (2.28) is given by the maximum principle and $f''(\eta) > 0$ for $\eta > 0$ (see Theorem 8.1 in [3]). We omit its details here (see [8]).

To show (2.4), from Corollary 1, we may consider the point $x = X_0 - d + \varepsilon$ as an initial position for sufficiently small $\varepsilon > 0$. Furthermore from (2.28) we obtain

$$\omega(X_0-d+\varepsilon,\psi) \leq \omega_B(\varepsilon,\psi) \leq m_B\psi\int_{X_0-d}^{X_0} dp_x(t)dt$$

for $0 \leq \psi \leq 2/m_B$, with $m_B >> 1$. Then the constant m with respect to the section $\omega(X_0-d+\varepsilon,\cdot)$, in the assumption for (2.20), can be chosen independently of any solution ω with $S(\omega_0) > X_0 - d$. Hence the constant A in (2.20) can be also chosen independently of $\omega(X_0-d+\varepsilon,\cdot)$. Then from Corollary 2 the inequality (2.4) holds. This proves Theorem 1.

3. Applications of Theorem 1.

3.1. Existence theorem of separation points for a more general exterior flow.

In this section we show the existence of separation points in the case where U(x) does not satisfy (2.1) and (2.2). We assume here that

(3.1) the exterior flow U(x) is sufficiently smooth and positive for $0 \leq x < \infty$

and

(3.2) the pressure gradient p_x is positive for some point $X_1 \in [0,\infty)$.

Let m be a positive constant such that

$$(3.3)\qquad \omega_0(\psi) \le \frac{m\psi}{\sqrt{X_0}} \int_0^{X_0} \bar{p}_x(t)dt \quad \text{for} \quad 0 \le \psi \le 2\sqrt{X_0}/m$$

where $\omega_0(\int_0^y u_0(t)dt) = u_0^2(y)$ for $u_0 \in I^{2+\alpha}$, the point X_0 and the function $\bar{p}_x$ will be defined below.

Theorem 2. For (1.1), (1.2) and (1.3) with the initial data $u_0 \in I^{2+\alpha}$, if the constant m in (3.3) is sufficiently small, then there exists a solution $(u,v) \in P^2([0,s))$ with the separation point $(s,0)$, $(0 < s < \infty)$.

Lemma 5. The pressure gradient p_x is monotone decreasing and positive on soem interval $[X_2,X_3] \subset [X_1,\infty)$.

Proof. Even if $p_x > 0$ on $[X_1,\infty)$, we find a sequence $\{x_n\}$ such that $p_x(x_n) \to 0$ as $n \to \infty$. In fact, assume that $p_x \ge a > 0$ on $[X_4,\infty)$ for some constant a and some point $X_4(>> 1)$. Then from (1.4) it implies that

$$\begin{aligned} U^2(x) &= U^2(X_4) + (x-X_4)\int_0^1 (U^2)_x(X_4+t(x-X_4))dt \\ &= U^2(X_4) - (x-X_4)\int_0^1 2p_x(X_4+t(x-X_4))dt \\ &\le U^2(X_4) - 2a(x-X_4) \\ &< 0 \quad \text{for } x >> 1. \end{aligned}$$

This is a contradiction. Hence, form the assumption of p_x, Lemma 5 is valid.

q.e.d.

From Lemma 5 and (1.4) we have that

$$U_x(X_3) < 0.$$

Let $\bar{U}(X)$ be the function such that

$$\bar{U}(x) = U(x) \quad \text{for} \quad x \leq X_3 \quad \text{and}$$

$$\bar{U}(x) = U(X_3) + (x-X_3)U_x(X_3) \quad \text{for} \quad x \geq X_3.$$

Furthermore set

$$\bar{p}_x(x) = -\bar{U}(x)\bar{U}_x(x) \quad \text{on} \quad [0,X_0],$$

where the point $x = X_0$ is the unique vanishing one of $\bar{U}(x)$. Then $\bar{p}_x$ is continuous, non-increasing, positive for $x \in [X_2,X_0)$ and $\bar{p}_x(X_0) = 0$.

Proof of Theorem 2. First we consider a solution $\omega(x,\psi) \in P_M^{2+\alpha}([0,S(\omega_0)))$ of (2.6), (2.7) and (2.8) with the exterior flow $\bar{U}(x)$, the pressure gradient $\bar{p}_x(x)$ and the initial data $\omega_0 \in I_M^{2+\alpha}$. Here we note that $\bar{U}$ and $\bar{p}_x \in C^0([0,X_0])$ satisfy (2.1) and (2.19) without the sufficiently smoothness. Now if we show

$$S(\omega_0) < X_3,$$

the solution $\omega(x,\psi)$ is the one of the original problem because of the definitions of $\bar{U}$ and $\bar{p}_x$.

To show this inequality for some ω_0, even if $S(\omega_0) > X_2$, let $\tilde{m}$ be a positive constant such that

$$\omega(X_2,\psi) \leq \frac{\tilde{m}\psi}{\sqrt{d}} \int_{X_2}^{X_0} \bar{p}_x(t)dt \quad \text{for} \quad 0 \leq \psi \leq 2\sqrt{d}/\tilde{m}$$

when $d = X_0 - X_2$.

Here we remark that

$$\int_x^{X_0} \bar{p}_x(t)dt = \int_x^{X_0} (-\bar{U}^2)_x(t)dt = \bar{U}^2(x) > 0$$

for $x \in [0,X_0)$ holds from the definitions of $\bar{p}_x$ and $\bar{U}$.
Then, if the constant m in (3.3) is sufficiently small, by the same way derived (2.23) it implies that the constant $\tilde{m}$ is also sufficiently small. Because (2.23) is valid without (2.19). Hence from (2.26) we see the desired a priori inequality for such an initial datum ω_0. Thus by the same way as in the proof of Theorem 1 we obtain Theorem 2.

q.e.d.

3.2. Approximation to the laminar Navier-Stokes flow.

For the sake of simplicity hereafter we are concerned with flows past a flat section of boundary. The stationary Navier-Stokes equations are

$$(3.4)\qquad \begin{aligned} uu_x + vu_y &= \nu(u_{xx} + u_{yy}) - p_x, \\ uv_x + vv_y &= \nu(v_{xx} + v_{yy}) - p_y, \\ u_x + v_y &= 0. \end{aligned}$$

The Prandtl equations (1.1) are formally derived from (3.4) by the following manner.

By the transformation

$$(3.5)\qquad \begin{aligned} &\eta = \nu^{-1/2}y, \quad x = x \quad \text{and} \\ &\tilde{u}(x,\eta) = u(x,y), \quad \tilde{v}(x,\eta) = \nu^{-1/2}v(x,y), \quad \tilde{p}(x,\eta) = p(x,y), \end{aligned}$$

we have

$$\tilde{u}_{\eta\eta} - \tilde{u}\tilde{u}_x - \tilde{v}\tilde{u}_\eta - \tilde{p}_x = -\nu\tilde{u}_{xx},$$

(3.6)
$$-\tilde{p}_\eta = -\nu(\tilde{v}_{\eta\eta} + \nu\tilde{v}_{xx}) + \nu\tilde{u}v_x + \nu\tilde{v}\tilde{v}_\eta,$$

$$\tilde{u}_x + \tilde{v}_\eta = 0.$$

Next, neglecting the order ν terms of the right sides in (3.6), we obtain the (dimensionless) Prandtl equations

(3.7)
$$\bar{u}_{\eta\eta} - \bar{u}\bar{u}_x - \bar{v}\bar{u}_\eta = \bar{p}_x(x),$$

$$\bar{u}_x + \bar{v}_\eta = 0,$$

where $\bar{p}_x(x) = \bar{p}_x(x,0)$. Then returning to the original coordinates (x,y) and denoting $(\bar{u}(x,\nu^{-1/2}y),\ \nu^{1/2}\bar{v}(x,\nu^{-1/2}y))$ by $(\bar{u}(x,y),\bar{v}(x,y))$ for convenience of discriptions, we obtain the Prandtl equations (1.1) for $(\bar{u}(x,y),\bar{v}(x,y))$.

In order to study the approximation results, considering all functions below to be continuous, for positive constants a (< 1), b_0 $(> a)$ and M (≥ 1) we define the calss Φ of *the laminar Navier-Stokes flows* (u^ν,v^ν,p^ν) *without separation points* in an interval $[0,A]$ (see [1] and its references).
The flow $(u^\nu,v^\nu,p^\nu) \in \Phi$ means the following :

(i) (u^ν,v^ν,p^ν) is the solution of (3.4) with the kinematic viscosity ν (≤ 1) in the domain $D = [0,A] \times [0,2]$, which satisfies the boundary conditions

$$u^\nu = v^\nu = 0 \quad \text{for} \quad y = 0.$$

(ii) $0 \leq u^\nu(x,y) \leq 1$ on D.

(iii) The absolute values of u^ν, v^ν and their x-derivatives up to order

2 are each bounded by M on D.

(iv) $|\nu(u^\nu_{xx} + u^\nu_{yy})|$, $|\nu(v^\nu_{xx} + v^\nu_{yy})|$ and $|\nu(v^\nu_{xx} + v^\nu_{yy})_x|$ are each bounded by M on D.

(v) $a\cdot\min\{1,\nu^{-1/2}y\} \le u^\nu(x,y) \le b_0\cdot\min\{1,\nu^{-1/2}y\}$ on $[0,A]\times[0,1]$.

Now for each $(u^\nu,v^\nu,p^\nu) \in \Phi$ we may take the sufficiently smooth pressure gradient $\bar{p}^\nu_x(x)$ in (3.7) and the initial data $\bar{u}^\nu_0(\eta) \in I^{2+\alpha}$ in (1.3) as follows :

For a positive constant B

$$(3.8)\qquad \begin{aligned} &|\tilde{p}^\nu_x(x,0) - \bar{p}^\nu_x(x)| \le B\nu^{1/2} \quad \text{if} \quad 0 \le x \le A, \\ &|\tilde{u}^\nu(0,\eta) - \bar{u}^\nu_0(\eta)| \le B\nu^{1/2}\min\{1,\eta\} \quad \text{if} \quad 0 \le \eta \le \nu^{-1/2} \quad \text{and} \end{aligned}$$

$$(3.9)\qquad \begin{aligned} &\bar{u}^\nu_0(\eta) \ge a\cdot\min\{1,\eta\} \quad \text{if} \quad 0 \le \eta \le \nu^{-1/2}, \\ &|\bar{p}^\nu_x(x)| \le b_1 \qquad\qquad \text{if} \quad 0 \le x \le A, \end{aligned}$$

provided that

$$(3.9)'\qquad \begin{aligned} &u^\nu(0,y) \quad \text{is monotonous and} \\ &|p^\nu_x(x,y)| \le b_1 \quad \text{on} \quad D \quad \text{for some positive constant } b_1. \end{aligned}$$

Finally, putting

$$(U^\nu)^2(x) \equiv (\bar{u}^\nu_0)^2(\infty) - 2\bar{p}^\nu(x) + 2\bar{p}^\nu(0),$$

we may take the exterior flow $U^\nu(x)$ as follows :

For some positive constant b_2 depending only on b_1

$$(3.10)\qquad 0 < b_2^{-1} \le (U^\nu)^2(x) \le b_2 \quad \text{if} \quad 0 \le x \le A.$$

Then setting $b = \max\{b_0, b_1, b_2\}$ and $B > b + 1$, we mention the following

Theorem 3. For the positive constants B, M, a and b let the flow $(u^\nu, v^\nu, p^\nu) \in \Phi$ and satisfy (3.9)'. Then, if $\nu \le \nu_0$ for a certain $\nu_0 = \nu_0(B,M,a,b)$, there exists the Prandtl flow $(\bar{u}^\nu, \bar{v}^\nu, \bar{p}^\nu)$ satisfying (3.8), (3.9) and (3.10) up to $x = A$ such that for a constant $c = c(B,M,a,b)$

$$(3.11)\qquad |u^\nu(x,y) - \bar{u}^\nu(x,y)| \le c\nu^{1/2} \quad \text{on} \quad [0,A] \times [0,\nu^{1/2}].$$

For convenience of the proof we consider the transformations (2.5) by virtue of the streaming functions $\tilde{\psi}(x,\eta)$ and $\bar{\psi}(x,\eta)$ of (3.6) and (3.7) respectively :

$$x = x \quad \text{and} \quad \tilde{\psi} = \tilde{\psi}(x,\eta) \quad \text{in} \quad (3.6),$$

$$x = x \quad \text{and} \quad \bar{\psi} = \bar{\psi}(x,\eta) \quad \text{in} \quad (3.7).$$

Then we have that $\omega^\nu(x,\tilde{\psi}) = (\tilde{u}^\nu)^2(x,\eta)$ and $\bar{\omega}^\nu(x,\bar{\psi}) = (\bar{u}^\nu)^2(x,\eta)$ satisfy

$$\sqrt{\omega^\nu}\, \omega^\nu_{\tilde{\psi}\tilde{\psi}} - \omega_x - 2\tilde{p}^\nu_x(x,0) = \nu^{1/2} f \quad \text{and}$$

$$\sqrt{\bar{\omega}^\nu}\, \bar{\omega}^\nu_{\bar{\psi}\bar{\psi}} - \bar{\omega}_x - 2\bar{p}^\nu_x(x) = 0,$$

where $\nu^{1/2} f(x,\eta) = 2\nu f_1 - 2\nu^{1/2}\int_0^\eta \partial f_2/\partial x \, d\eta'$, $\nu f_1(x,\eta) = \nu\tilde{u}^\nu_{xx}$ and $\nu^{1/2} f_2(x,\eta) = -\nu(\tilde{v}^\nu_{\eta\eta} + \nu\tilde{v}_{xx}) + \nu\tilde{u}^\nu\tilde{v}^\nu_x + \nu\tilde{v}^\nu\tilde{v}^\nu_\eta$.

The outline of the proof of Theorem 3. From the assumptions mentioned above and by the Fife's theory ([1]) we need only to prove the following: For any $\nu \leq \nu_0 = \nu_0(B,M,\frac{\tilde{a}}{3},b)$ there exists a Prandtl flow $(\bar{u}^\nu,\bar{v}^\nu,\bar{p}^\nu) \in P^2([0,A])$ such that $\bar{\omega}^\nu(x,\bar{\psi}) \geq \frac{\tilde{a}}{3}\bar{\psi}$ on $[0,A] \times [0,1]$, where $\tilde{a} = \frac{2a^2}{b_0+B}$.

Now in order to obtain the continuation $\bar{\omega}^\nu$ mentioned above up to $x = A$, we put $A_\nu = \sup\{\ell$; for any $\nu' \leq \nu$ the solution $\bar{\omega}^{\nu'}(x,\bar{\psi})$ exists

in $P_M^{2+\alpha}([0,\ell])$ and satisfies

$$\bar{\omega}^{\nu'}(x,\bar{\psi}) \geq \frac{\tilde{a}}{3}\bar{\psi} \tag{3.12}$$

on $[0,\ell] \times [0,1]\}$.

Obviously from the proof of Lemma 4 (in particular see the one of Lemma 2 in [11]) it implies that $A_1 > 0$. Assume that $A_1 < A$. Let ℓ be a point such that $\ell < A_1$ and $A_1 - \ell << 1$. Then from (3.12), all assumptions (i),$\cdots$,(v), (3.8),$\cdots$, (3.10) and by the analogous way as in the proof of Lemma 5.3 in [1] we obtain that for a certain positive constant $\bar{c}_1 = \bar{c}_1(B,M,\frac{\tilde{a}}{3},b)$

$$|\omega^\nu(x,\bar{\psi}) - \bar{\omega}^\nu(x,\bar{\psi})| \leq \bar{c}_1\nu^{1/2}\bar{\psi}$$

if $(x,\bar{\psi}) \in [0,\ell] \times [0,1]$ and $\nu \leq 1$.

Therefore from (v) in Φ we have that for $\nu \leq \min\{(\frac{\tilde{a}}{2\bar{c}_1})^2,1\} \equiv \nu_0$

$$\bar{\omega}^\nu(x,\bar{\psi}) \geq \frac{\tilde{a}}{2}\bar{\psi} \quad \text{on} \quad [0,\ell] \times [0,1].$$

Accordingly from (3.9), (3.9)', (3.10) and by using Corollary 1 and the proof of Lemma 4 we conclude that for such ν $\bar{\omega}^\nu$ can be continued and satisfies (3.12) up to $x = \ell + \delta$, where the positive constant δ does not depend on ℓ. Thus we see that $A_{\nu_0} > A_1$. Furthermore by the same way as above we may conclude that $A_{\nu_0} = A$.

Finally we must mention the following : ν_0 in Theorem 3 tends to zero rapidly if a tends to zero and if M, B, b are fixed. Therefore in order to keep ν_0 not too small, we must assume that $1 > a >> 0$.

References

[1] Fife, P.C.: Considertations regarding the mathematical basis for Prandtl's boundary layer theory, Arch. Rat. Mech. Anal., 28 (1968), 184-216.

________ : Corrigendum, Considerations regarding the mathematical basis for Prandtl's boundary layer theory, Arch. Rat. Mech. Anal., 46 (1972), 389-393.

[2] Glimm, J.: Singularities in fluid dynamics, Math. Prob. in Theoretical Phys., ed. R. Schrader, R. Seiler and D.A. Uhlenbrock, Springer-Verlag (1981), 86-97.

[3] Hartman, P.: Ordinary Differential Equations, John Wiley and Sons Inc., New York (1964).

[4] Hastings, S.P.: Reversed flow solutions of the Falkner-Skan equations, SIAM. Appl. Math. Vol. 22, No.2 (1972), 329-334.

[5] Il'in, A.M., Kalashnikov, A.S. and Oleinik, O.A.: Linear equations of the second order of parabolic type, Russian Math. Survey, 17-3 (1962), 3-146.

*[6] Lagerstorm, P.A.: Solutions of Navier-Stokes equations at large Reynolds number, SIAM. Appl. Math. Vol. 28, No.1 (1975), 202-214.

*[7] Landau, L.D. and Lifshitz, E.M.: Fluid Mechanics, Pergamon press,

Oxford (1966).

[8] Matsui, S. and Shirota, T.: On separation points of solutions to Prandtl boundary layer problems, Hokkaido Math. Jour. Vol. 13, No.1 (1984), 92-108.

*[9] Von Mises, R. and Friedrichs, K.O.: Fluid Dynamics, Appl. Math Sciences 5, Springer-Verlag (1971).

[10] Nickel, K.: Parabolic equations with applications to boundary layer theory, P. D. E. and Conti. Mech., ed. R. Langer, The Univ. Wisconsin Press, Madison, Wisconsin (1961), 319-330.

[11] Oleinik, O.A.: On a system of equations in boundary layer theory, U. S. S. R. Comp. Math. Phys., 3 (1963), 650-673.

[12] __________ : Mathematical problems of boundary layer theory, Uspehi Mat. Nauk, Vol. 23, No. 3 (1968), 3-65.

[13] __________ : Weak solutions in the Sobolev sense for a system of boundary layer equations, Amer. Math. Soc. Trans. (2), 105 (1976), 247-264.

[14] _____ and Kruzhkov, S.N.: Quasi-linear second-order parabolic equations with many independent variables, Russian Math. Surveys, Vol. 16 n.5 (1961), 106-146.

[15] Serrin, J.: Asymptotic behavior of velocity profiles in the Prandtl boundary layer theory, Proc. London Math. Soc. A 299 (1967), 491-507.

*[16] Williams, III, J.C.: Incompressible boundary-layer separation, Ann. Rev. Fluid Mech., 9 (1977), 113-144.

An asteriks * is used to mark the references developing the theory of Fluid Mechanics in physical and enginering points of view.

This Page Intentionally Left Blank

Lecture Notes in Num. Appl. Anal., 8, 107–125 (1985)
Recent Topics in Nonlinear PDE II, Sendai, 1985

On a Free Boundary Problem in Ecology

Masayasu MIMURA*, Yoshio YAMADA** and Shoji YOTSUTANI***

*Department of Mathematics, Hiroshima University
Hiroshima 730, Japan

**Department of Mathematics, Nagoya University
Nagoya 464, Japan

***Department of Applied Science, Miyazaki University
Miyazaki 880, Japan

We shall be concerned with a free boundary problem for semilinear parabolic equations, which describes the habitat segregation phenomenon in population ecology. The main purpose is to show the global existence, uniqueness, regularity and asymptotic behavior of solutions for the problem. The stability or instability of each stationary solution is completely determined using the comparison principle.

§1. Problem

We consider the following one-dimensional free boundary problem.

$$(\mathrm{P})\begin{cases} (1.1) & u_t = d_1 u_{xx} + f(u)u \quad \text{in } S^-, \\ (1.2) & u_t = d_2 u_{xx} + g(u)u \quad \text{in } S^+, \\ (1.3) & u(0,t) = m_1, \quad t \in (0,\infty), \\ (1.4) & u(1,t) = -m_2, \quad t \in (0,\infty), \\ (1.5) & u(s(t),t) = 0, \quad t \in (0,\infty), \\ (1.6) & u(x,0) = \varphi(x), \quad x \in I, \\ (1.7) & \dot{s}(t) = -\mu_1 u_x(s(t)-0,t) + \mu_2 u_x(s(t)+0,t), \ t \in (0,\infty), \\ (1.8) & s(0) = \ell, \end{cases}$$

where S^- (resp. S^+) is the open set of $Q = I \times (0,\infty)$ ($I = (0,1)$) in which $x < s(t)$ (resp. $x > s(t)$), and $u_x(s(t)-0,t)$ (resp. $u_x(s(t)+0,t)$) denotes the limit of $u(x,t)$ at $x = s(t)$ from the left (resp. right).

This is a model describing regional partition of two species, which are struggling on a boundary to obtain their own habitats. In our model, the function u in S^- (resp. $-u$ in S^+) denote the population density of the species which lives in S^- (resp. S^+). These two spieces are supposed to undergo diffusion and growth. Here the boundaries $x = 0, 1$ are fixed, while the intermidiate boundary $x = s(t)$ is not prescribed a priori. The latter boundary is determined by the interactions between the two species there (see (1.5) and (1.7)). In this sense, $x = s(t)$ is called a free boundary. For details, see [3].

In the absense of f and g, the problem is reduced to a two phase Stefan problem in the one-dimensional space, for which there are many contributions (see Rubinstein [5], Kamenomostskaja [2], Yamaguti & Nogi [6], Yotsutani [7] e.t.c., and references therin). We shall show several interesting results on the global existence, uniqueness, regularity and asymptotic behavior of solutions of (P). Especially, a bifurcation phenomenon occurs in (P), differently from the one in the case $f = g = 0$ (Stefan problem).

§2. Assumption

We summerize the assumptions:

(A.1) f is locally Lipschitz continuous on $[0,\infty)$, monotone non-increasing on $[0,1]$ and satisfies

$$f(u) > 0 \quad \text{on} \quad [0,1), \quad f(1) = 0 \quad \text{and} \quad f(u) \leq 0 \quad \text{on} \quad (1,\infty).$$

(A.2) g is locally Lipschitz continuous on $(-\infty,0]$, monotone non-decreasing on $[-1,0]$ and satisfies

$$g(u) > 0 \quad \text{on} \quad (-1,0], \quad g(-1)=0 \quad \text{and} \quad g(u) \leq 0 \quad \text{on} \quad (-\infty,-1).$$

(A.3) $0 < m_1 \leq 1$ and $0 < m_2 \leq 1$.

(A.4) $0 < \ell < 1$.

(A.5) $\phi \in H^1(I)$ satisfies $\phi(0) = m_1$, $\phi(1) = -m_2$ and

$$(\ell - x)\phi(x) \geq 0, \quad \text{for} \quad x \in I.$$

Our first result is concerned with the global solvability of (P).

<u>Theorem I</u>. (Global existence and uniqueness). Under the assumptions (A.1)-(A.5), there exists a unique pair of functions $(u, s) \in C(\overline{Q}) \times C([0,\infty))$ satisfying the following properties:

(i) $s(0) = \ell$, $s \in L^3(0,\infty)$ and for some constant $b \in (0,1)$

$$b \leq s(t) \leq 1 - b, \quad t \in [0,\infty).$$

(ii) u satisfies (1.3), (1.4), (1.5) and (1.6) everywhere and

$$0 \leq u(x,t) \leq \overline{M} = \max\,(1, \sup_{0\leq x\leq \ell} \phi(x)) \quad \text{in} \quad \overline{S^-},$$

$$0 \geq u(x,t) \geq -\overline{M} = \min\,(-1, \inf_{\ell\leq x\leq 1} \phi(x)) \quad \text{in} \quad \overline{S^+}.$$

(iii) Let $u^+ = \max\,(u, 0)$ and $u^- = -\min\,(u,0)$. Then

$u^{\pm} \in C([0,\infty);H^1(I))$ and

$$\sup_{0\leq t<\infty} \|u^+(t)\|_{H^1(I)} < \infty, \quad \sup_{0\leq t<\infty} \|u^-(t)\|_{H^1(I)} < \infty.$$

(iv) $u_t \in L^2(S^-) \cap L^2(S^+)$.

(v) u_t, $u_{xx} \in C(S^-) \cap C(S^+)$ and u satisfies (1.1) and (1.2) everywhere in S^- and S^+, respectively.

(vi) For each $\delta > 0$, u_x is Hölder continuous in $(x,t) \in \overline{S^{\pm}} \cap \{t \geq \delta\}$ and $\dot{s}$ is Hörder continuous in $t \in [\delta,\infty)$.

(vii) $\{u, s\}$ satisfies (1.7) for every $t \in (\delta,\infty)$.

In this report, we say that $\{u, s\}$ is a <u>smooth</u> <u>solution</u> of (P) on $[0,\infty)$ if it has the properties in Theorem I. Let $\{u(x,t;\phi,\ell), s(t;\phi,\ell)\}$ denote the smooth solution of (P) with initial data $\{\phi, \ell\}$. Theorem I assures the global existence of the smooth solution $\{u(x,t;\phi,\ell), s(t;\phi,\ell)\}$ for every $\{\phi, \ell\}$, so that its asymptotic behavior as $t \to \infty$ is now to be discussed. For this purpose, we conviniently introduce the <u>ω-limit</u> <u>set</u> defined by

$$\omega(\phi,\ell) = \{\{u^*, s^*\} \in H^1(I) \times I;\ \text{there exists a sequence } \{t_i\} \to \infty \text{ such that } s(t_i;\phi,\ell) \to s^*,\ u(\cdot,t_i;\phi,\ell) \to u^* \text{ in } C(\bar{I}) \text{ and } u^{\pm}(\cdot,t_i;\phi,\ell) \to (u^*)^{\pm} \text{ in } H^1(I) \text{ as } i \to \infty\}. \tag{2.1}$$

We say that the sequence $\{u(\cdot,t_i;\phi,\ell), s(t_i;\phi,\ell)\}$ converges to $\{u^*, s^*\}$ in <u>Ω-topology</u> if it has the convergence properties in (2.1)

<u>Theorem II</u>. (structure of ω-limit set). Assume (A.1)-(A.5).

Then

(i) $\omega(\varphi,\ell)$ is non-empty and connected in Ω-topology.

(ii) If $(u^*, s^*) \in \omega(\varphi,\ell)$, then (u^*, s^*) satisfies

$$
\text{(S.P)} \quad \begin{cases} d_1 u^*_{xx} + f(u^*)u^* = 0, \quad u^* \geq 0, & x \in (0,s^*), \\ d_2 u^*_{xx} + g(u^*)u^* = 0, \quad u^* \leq 0, & x \in (s^*,1), \\ u^*(0) = m_1, \quad u^*(s^*) = 0, \quad u^*(1) = -m_2, \\ -\mu_1 u^*_x(s^*-0) + \mu_2 u^*_x(s^*+0) = 0. \end{cases}
$$

Theorem II gives very useful information about the asymptotic behavior of solutions of (P). For example, if one can show that solutions of (SP) are isolated in Ω-topology, then $(u(t;\varphi,\ell), s(t;\varphi,\ell))$ approaches one of them in Ω-topology as $t \to \infty$. (The limiting function will depend on the initial data.) Therefore, it is very important to determine the structure of the set of solutions for (SP).

We now study the stationary problem (SP) with the aid of the following auxiliary problem

$$
(2.2) \quad \begin{cases} d_1 w_{xx} + wf(w) = 0, \quad w \geq 0, & x \in (0,\xi), \\ d_2 w_{xx} + wg(w) = 0, \quad w \leq 0, & x \in (\xi,1), \\ w(0) = m_1, \quad w(\xi) = 0, \quad w(1) = -m_2, \end{cases}
$$

where $\xi \in I$ is any fixed number.

Then we have

Theorem III. (Stationary solutions). Assume (A.1)-(A.3).

Then

(i) For every $\xi \in I$, (2.2) has a unique solution $w = w(x;\xi)$.

(ii) $(w(x;s^*), s^*)$ is a solution of (SP), if and only if s^* is a zero point of

$$W(\xi) = -\mu_1 w_x(\xi-0;\xi) + \mu_2 w_x(\xi+0;\xi). \tag{2.3}$$

(iii) If $\xi_1 < \xi_2$, then $w(x;\xi_1) < w(x;\xi_2)$ in I.

(iv) (SP) has a (unique) minimal solution $(\underline{u}, \underline{s}) = (w(\cdot;\underline{s}), \underline{s})$ and a (unique) maximal solution $(\bar{u}, \bar{s}) = (w(\cdot,\bar{s}), \bar{s})$ in the sense that any solution (u^*, s^*) of (SP) satisfies

$\underline{s} \leq s^* \leq \bar{s}$ and $\underline{u} \leq u^* \leq \bar{u}$ in $\bar{I}$.

Here $\underline{s}$ and $\bar{s}$ are the least zero point and the greatest zero point of W on I, respectively.

By the properties (iii) of Theorem III, the set of solutions for (SP) has an obvious order relation. For $(u_i, s_i) \in C(\bar{I}) \times \bar{I}$ ($i = 1, 2$), we write

$$(u_1, s_1) \leq (u_2, s_2),$$

if

$$u_1 \le u_2 \quad \text{on} \quad \bar{I}, \qquad s_1 \le s_2.$$

We are now in a position to study stability or instability of each solution of (SP) in connection with the asymptotic behavior of solutions for (P).

<u>Theorem IV</u>. (Stability of stationary solutions). Assume (A.1)-(A.5). Then

(i) The minimal solution $\{\underline{u}, \underline{s}\}$ of (SP) is globally aymptotically stable from below in the sense that if

$$\{\phi, \ell\} \le \{\underline{u}, \underline{s}\},$$

then the smooth solution $\{u(t;\phi,\ell), s(t;\phi,\ell)\}$ of (P) satisfies

$$\{u(t;\phi,\ell), s(t;\phi,\ell)\} \le \{\underline{u}, \underline{s}\} \qquad \text{for every} \quad t \in [0,\infty),$$

and

$$(2.4) \quad \lim_{t\to\infty} \{u(t;\phi,\ell), s(t;\phi,\ell)\} = \{\underline{u}, \underline{s}\} \quad \text{in } \Omega\text{-topology.}$$

(ii) The maximal solution $\{\bar{u}, \bar{s}\}$ of (SP) is globally asymptotically satable from above, that is, the assertion of (i) is valid if $\{\underline{u}, \underline{s}\}$ and $\le$ are replaced by $\{\bar{u}, \bar{s}\}$ and $\ge$, respectively.

(iii) Let $\xi_1 < \xi_2$ be adjacent zero points of $W(\xi)$ defined by (2.3). If

$$(2.5)\quad (w(\cdot;\xi_1),\ \xi_1) \le (\phi,\ \ell) \le (w(\cdot,\xi_2),\ \xi_2).$$

then

$$(w(\cdot,\xi_1),\ \xi_1) \le (u(t;\phi,\ell),\ s(t;\phi,\ell)) \le (w(\cdot,\xi_2),\ \xi_2)$$

for every $t \in [0,\infty)$.

Moreover, if $W(\xi) > 0$ (resp. $W(\xi) < 0$) for $\xi \in (\xi_1,\xi_2)$, then $(w(\cdot,\xi_2),\ \xi_2)$ (resp. $(w(\cdot,\xi_1),\xi_1)$) is asymptotically stable from below (resp. from above) in the sense that, for any $(\phi,\ \ell)$ satisfying (2.5) with $w(\cdot,\xi_1) \not\equiv \phi$ on $\bar{I}$ (resp. $w(\cdot,\xi_2) \not\equiv \phi$ on $\bar{I}$), the convergence property (2.4) holds with $(\underline{u},\ \underline{s})$ replaced by $(w(\cdot,\xi_2),\ \xi_2)$ (resp. $(w(\cdot,\xi_1),\ \xi_1)$).

Remark 2.1. As a special case, we take

$$d_1 = d_2 = d,\quad \mu_1 = \mu_2 = \mu,\quad m_1 = m_2 = m,$$

and

$$f(u) = g(-u) = \lambda h(u),$$

where h is a function satisfying (A.1).

We define

$$\Lambda(m,d,\lambda) = \left(\frac{d}{2\lambda}\right)^{1/2} \int_0^m (\ H(m) - H(u)\ )^{-1/2}\, du$$

with

$$H(u) = \int_0^u v\, h(v)\, dv.$$

After some calculations, it is proved that $W(\xi)$ has only one zero point $\xi = 1/2$ if $\Lambda \geq 1/2$ and that it has three zero points $\xi = s', 1/2, 1-s'$ with $s' \in (0,1/2)$ if $\Lambda < 1/2$. Therefore Theorem III implies that, if $\Lambda \geq 1/2$, then (SP) has a unique solution $\{u_s, 1/2\}$, which is globally asymptoyically stable from below and above by Theorem IV. In the case $\Lambda < 1/2$, (SP) has three solutions $\{\underline{u}, s'\}$, $\{u_s, 1/2\}$ and $\{\bar{u}, 1-s'\}$, of which $\{\underline{u}, s'\}$ and $\{\bar{u}, 1-s'\}$ are asymptotically stable from below and above by Theorem IV. If we take, say, λ as a parameter, these cosiderations exhibit a bifurcation phenomenon. If λ is smaller than a critical value λ_0 (recall the definition of Λ), every solution of (P) converges to $\{u_s, 1/2\}$. However, if λ becomes greater than λ_0, then $\{u_s, 1/2\}$ loses its sability and two bifurcated solutions $\{\underline{u}, s'\}$ and $\{\bar{u}, 1-s'\}$ get the the asymptotic stability.

<u>Example 2.1</u>. Suppose that

$$f(u) = g(-u) = \lambda\, (1 - u),$$

$$d_1 = d_2 = 1, \quad \mu_1 = \mu_2 = 1 \quad , \; m_1 = m_2 = \tfrac{1}{2}.$$

Then (SP) exhibits bifurcation phenomenon when $\lambda > 0$ is a bifurcation parameter (see Figures 1, 2, and 3).

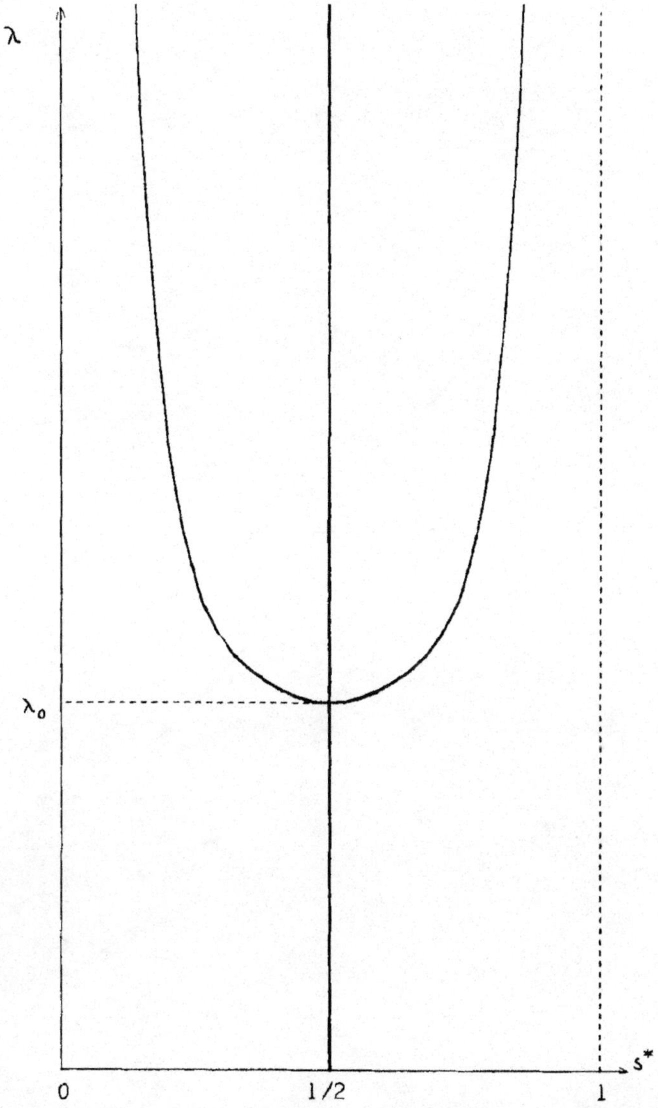

Figure 1. The bifurcation diagram of s^* in (SP) with respect to λ.

$$\lambda_0 = 4 \left[\int_0^{1/2} \left(\frac{1}{6} - u^2 + \frac{2}{3} u^3 \right)^{-1/2} du \right]^2$$

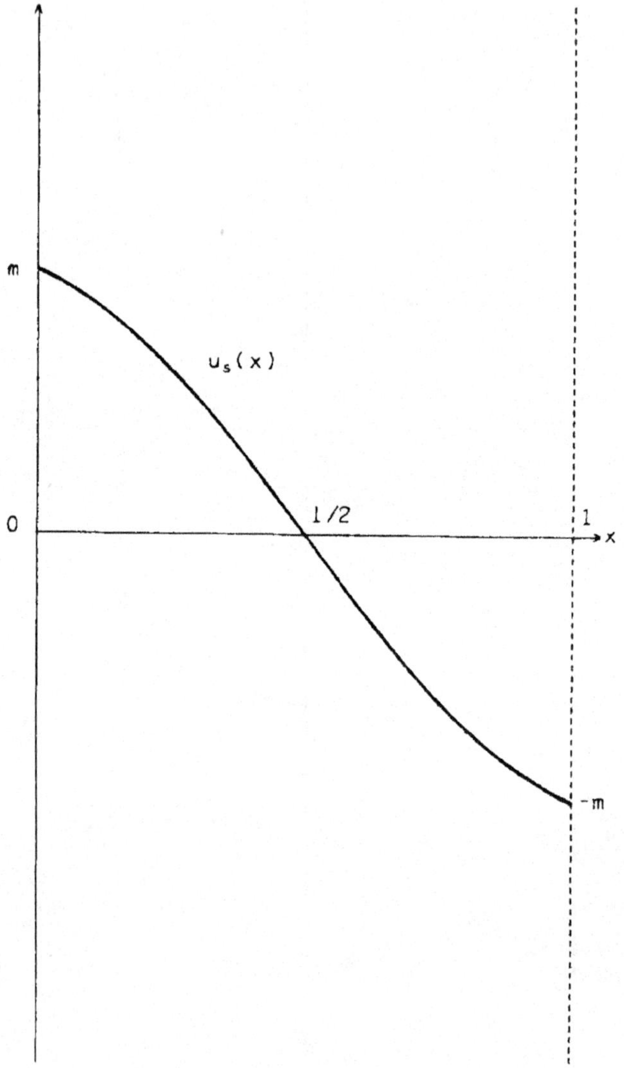

Figure 2. Profile of $u^*(x)$ with $\lambda = 10$

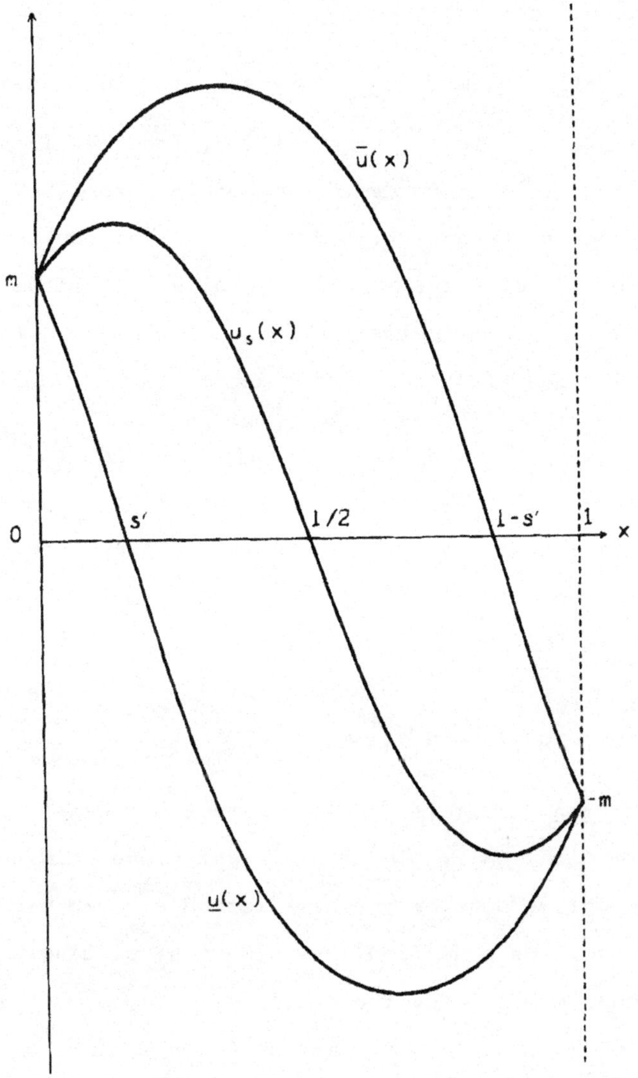

Figure 3. Profile of $u^*(x)$ with $\lambda = 40$

§3. Concluding remarks

Remark 3.1. Even if we drop the monotonicity assumption of f (resp. g) in (A.1) (resp. (A.2)), Theorems I and II remain true. However, the uniqueness of solutuions for (2.2) is not assured, so that the analysis of (SP) will become more complicated. However the stability analysis of solutions of (SP) can be carried out along the line developed in this report (for details, see [4]).

Remark 3.2. In the case where $m_1 = 0$ or $m_2 = 0$, the free boundary $x = s(t)$ may hit one of the fixed boundaries $x = 0, 1$ in a finite time. Therefore, we need more careful analysis to get complete information about the behavior of solutions for (P) (for details, see [4]).

Similar situation occurs when (1.3) and (1.4) are replaced by Neumann boundary conditions.

Remark 3.3. In this report, we have discussed a one-dimensional model only. The analysis of higher dimensional models can not be done by the same method as used here. This is still open. Here we display a numerical simulation of a 2-dimensional model where the region K is the unit square $\{(x,y) ; x \in I, y \in I\}$ (see Figures 4 - 8). The equations are described by the 2-dimensional version of (1.1)-(1.8), where

$$f(u) = g(-u) = \lambda\ (1 - u), \quad d_1 = d_2 = 1, \quad \mu_1 = \mu_2 = 1.$$

The boundary conditions are

$$u(0,y,t) = 0.7, \quad y \in I, \quad t > 0,$$

$$u(1,y,t) = 0.7, \quad y \in I, \quad t > 0,$$

$$u_y(x,0,t) = 0, \quad x \in I, \quad t > 0,$$

$$u_y(x,1,t) = 0. \quad x \in I, \quad t > 0.$$

It seems that 2-dimensional steady state solutions uniformly in $y \in I$ (which are essencially the one-dimensional solutions obtained in the above) are also stable in 2-dimensional problems. We note that the papers [1], [8] and the references therein are worthly of attention.

Test 3 (Initial distribution)

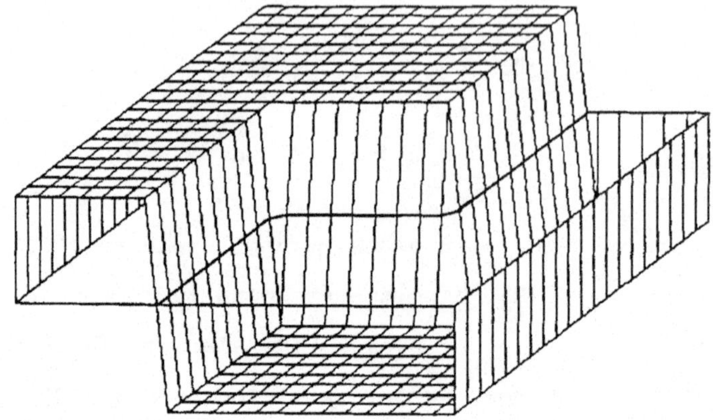

C1: 0.1000E 1 C2: 0.1000E 1 G1: 0.1000E 1 G2: 0.1000E 1
BCM: 0.700E 0 BCN: 0.700E 0

Figure. 4

Test 3 ($t = 0.02$)

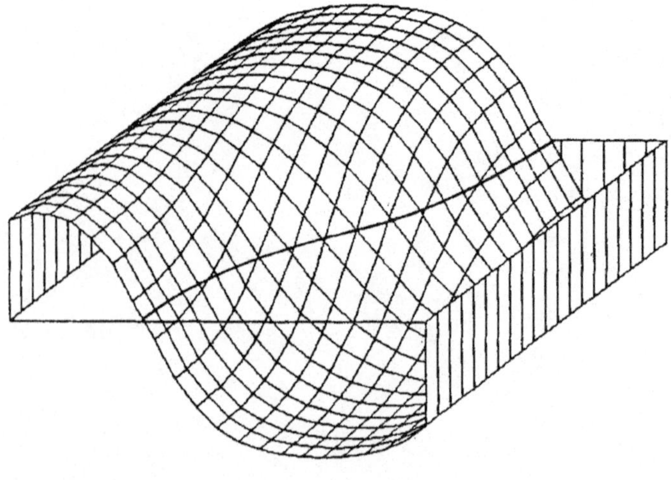

C1: 0.1000E 1 C2: 0.1000E 1 G1: 0.1000E 3 G2: 0.1000E 3
BCM: 0.700E 0 BCN: 0.700E 0

Figure. 5

Test 3 (t = 0.2)

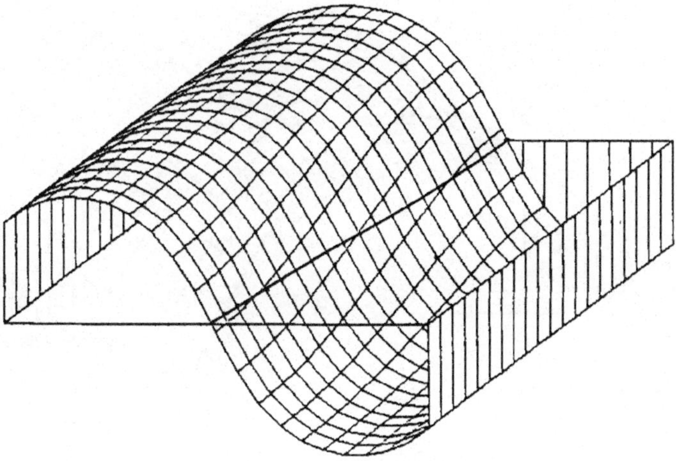

C1: 0.1000E 1 C2: 0.1000E 1 G1: 0.1000E 3 G2: 0.1000E 3
BCM: 0.700E 0 BCN: 0.700E 0

Figure. 6

Test 3 (t = 1.)

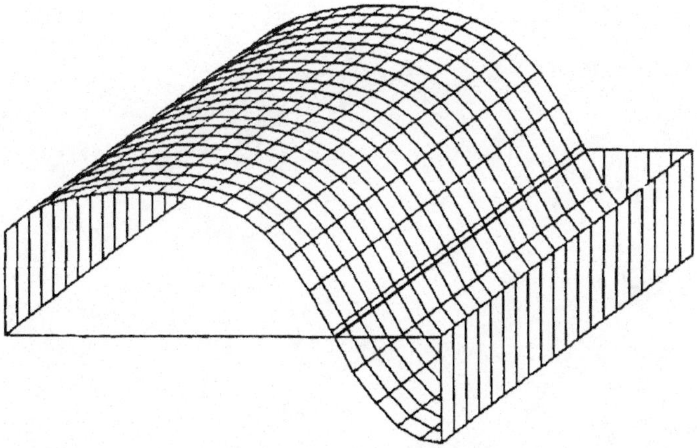

C1: 0.1000E 1 C2: 0.1000E 1 G1: 0.1000E 3 G2: 0.1000E 3
BCM: 0.700E 0 BCN: 0.700E 0

Figure. 7

Test 3 (t = 4.)

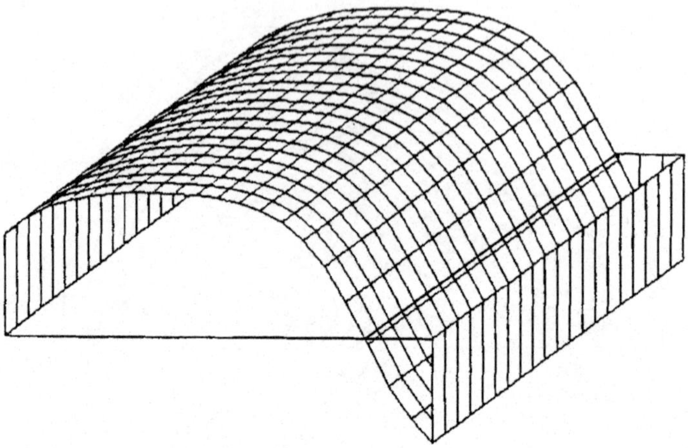

C1: 0.1000E 1 C2: 0.1000E 1 G1: 0.1000E 3 G2: 0.1000E 3
BCM: 0.700E 0 BCN: 0.700E 0

Figure. 8

References

[1] R. K. Alexander and B.A. Fleishman. Perturbation and Bifurcation in a Free Boundary Problem. J. D. E., 45 (1982), 34-52.

[2] S. L. Kamenomostskaja. On Stefan's problem. Mat. Sb., 53 (1961), 489-514.

[3] M. Mimura, Y. Yamada and S. Yotsutani. A free boundary problem in ecology I. To appear in J.J.A.M.

[4] M. Mimura, Y. Yamada and S. Yotsutani. A free boundary problem in ecology II. In preparation.

[5] L. I. Rubinstein. The Stefan Problem. Translation of Mathematical Monographs 27. Amer. Math. Soc. providience, R. I., 1971.

[6] M. Yamaguti and T. Nogi. The Stefan problem. Sangyo-Tosho, 1977, (in Japanese).

[7] S. Yotsutani. Stefan problems with the unilateral boundary condition on the fixed boudary IV. Osaka J. Math., 21 (1984), 149-167

[8] D. Aronson, M. G. Crandall and L. A. Peletier. Stabilization of solutions of a degenerate nonlinear problem. Nonlinear Analysis , 6 (1982), 1001-1022.

This Page Intentionally Left Blank

Lecture Notes in Num. Appl. Anal., 8, 127–132 (1985)
Recent Topics in Nonlinear PDE II, Sendai, 1985

On the Vanishing Viscousity of the Incompressible Fluid in the Whole Plane

Ryûichi Mizumzchi

Mathematical Institution, Tôhoku University
Sendai 980, Japan

Many works are done as to the convergence of the solutions of the Navier-Stokes equations (NS) in space domain R^2 to those of the Euler equations (E) as the viscousity parameter ν goes to 0 ([1]-[4]).

$$
(NS)\quad \begin{cases} \dfrac{\partial}{\partial t} u^{(\nu)} - \nu\Delta u^{(\nu)} + (u^{(\nu)}\cdot\nabla)u^{(\nu)} + \nabla p^{(\nu)} = 0 \\ \operatorname{div} u^{(\nu)} = 0 \\ u^{(\nu)}|_{t=0} = u_0 \end{cases}
$$

$$
(E)\quad \begin{cases} \dfrac{\partial}{\partial t} u^{(0)} + (u^{(0)}\cdot\nabla)u^{(0)} + \nabla p^{(0)} = 0 \\ \operatorname{div} u^{(0)} = 0 \\ u^{(0)}|_{t=0} = u_0 \end{cases}
$$

Among others, a recent paper of T. Kato [2] showed that if $u_0 \in H^s(R^2)$ for $s>2$, then there are unique solutions $u^{(\nu)}$ and $u^{(0)}$ to (NS) and (E), respectively, in $C([0,\infty);\ H^s)$ and $u^{(\nu)} \to u^{(0)}$ in $C([0,T);\ H^s)$ as $\nu \to 0$, for any $T>0$. (The same result has been obtained by Kyûya Masuda in an unpublished paper.) Here we discuss this problem in Hölder spaces.

Let T be arbitrarily fixed. We first construct solutions of (NS) and (E) in the domain $(0,T)\times R^2$, referring to [3]. Let F_1 be the operator defined by

$$F_1\omega(t,x) = \frac{1}{2\pi}\iint_{R^2} \frac{(x-y)^{\perp}}{|x-y|^2}\,\omega(t,y)\,dy_1dy_2$$

where $x^{\perp}=(x_2,-x_1)$ for $x=(x_1,x_2)$. For $\nu\geq 0$, let us define $F_2^{(\nu)}$, an operator from $\{u\in(C([0,T)\times R^2)^2 : u(t,\cdot)\in(C^1(R^2))^2$ and $\operatorname{div} u(t,\cdot)=0$, $0\leq t<T\}$ to $C([0,T)\times R^2)$, by $F_2^{(\nu)}u=\omega$, where ω is the unique bounded solution of the following equation (1) with a given bounded continuous initial data ω_0:

$$(1)\quad \begin{cases} \dfrac{\partial}{\partial t}\omega - \nu\Delta\omega + (u\cdot\nabla)\omega = 0 \\ \omega|_{t=0} = \omega_0. \end{cases}$$

Let $F^{(\nu)}=F_2^{(\nu)}F_1$. Let $B_\alpha=C^\alpha(R^2)\cap L^1(R^2)$, $0<\alpha<1$, equipped with the norm $|f|_{B_\alpha}=|f|_{L^1}+|f|_{L^\infty}+\frac{1}{\alpha}\sup_{x,y\in R^2}\frac{|f(x)-f(y)|}{|x-y|^\alpha}$, and let $X_\alpha=C([0,T)\times R^2)$ $\cap L^\infty((0,T);B_\alpha)$ with the norm $|g|_{X_\alpha}=\sup_{0<t<T}|g(t,\cdot)|_{B_\alpha}$, $g\in X_\alpha$. Then, if $\omega_0=\operatorname{rot} u_0$ and ω is a fixed point of $F^{(\nu)}$ in X_α, then $F_1\omega$ is a solution of (NS) or (E) (see [3]). Unique existence of the fixed point of $F^{(\nu)}$ is established by the following theorem.

Theorem 1. Suppose $\omega_0\in B_\alpha$. Then there is a unique fixed point of $F^{(\nu)}$ in X_α to each $\nu\geq 0$. These fixed points are uniformly bounded in X_α for small ν.

Outline of proof. To construct the solution ω of (1) let us prepare some notations. For $x \in R^2$ and $0 \leq s,t<T$, $U_{t,s}(x)$ is defined as the solution of the differential equation

$$(2)\begin{cases} \frac{d}{dt} U_{t,s}(x) = u(t,U_{t,s}(x)) \\ U_{s,s}(x) = x. \end{cases}$$

A function Z, integral operators W, Φ and Ψ, and a differential operator L are given by

$$Z(t,x;\ s,y) = \frac{1}{4\pi\nu(t-s)} \exp\left(- \frac{|x-U_{t,s}(y)|^2}{4\nu(t-s)} \right)$$

$$W[f](t,x) = \iint_{R^2} Z(t,x;\ 0,y)f(y)\ dy_1 dy_2$$

$$\Phi[g](t,x) = \int_0^t \iint_{R^2} Z(t,x;\ s,y)g(s,y)\ dy_1 dy_2\ ds$$

$$\Psi[g](t,x) = \int_0^t \iint_{R^2} LZ(t,x;\ s,y)g(s,y)\ dy_1 dy_2\ ds$$

$$L = - \frac{\partial}{\partial t} + \nu\Delta_x - (u(t,x)\cdot\nabla_x).$$

Then the solution ω of (1) can be given by

$$(3) \qquad \omega = W[\omega_0] + \sum_{j=0}^{\infty} \Phi\Psi^j LW[\omega_0].$$

We note the following lemma due to McGrath [3].

Lemma 1 (McGrath). Suppose $\zeta \in L^\infty((0,T);\ L^1(R^2)\cap L^\infty(R^2))$, $u=F_1\zeta$ and $U_{t,s}(x)$ is the solution of (2). Then there exist δ and M, depending only on T and $\sup_{0<t<T}(|\zeta(t,\cdot)|_{L^1} + |\zeta(t,\cdot)|_{L^\infty})$, such that

$$|U_{t,s}(x) - U_{t,s}(y)| \leq M|x-y|^{\delta}, \quad x,y\in R^2, \quad 0\leq s,t<T.$$

The following lemma is essential to prove theorem 1.

Lemma 2. Suppose $\zeta\in X_{k'}$, $k'>0$, and $u=F_1\zeta$. Let $0<h\leq k<1$. Then

$$\sup_{0\leq t<T} |\Psi[g](t,\cdot)|_{C^h(R^2)} \leq c\sqrt{\nu}^{\,k-h} |\zeta|_{X_{k'}} \exp(c|\zeta|_{X_{k'}})$$

$$\cdot \sup_{0\leq t<T} |g(t,\cdot)|_{C^k(R^2)}$$

where $|\zeta|_{C^\gamma(R^2)} = \sup_{x,y\in R^2, x\neq y} \dfrac{|\zeta(x)-\zeta(y)|}{|x-y|^\gamma}$ for $\gamma>0$ and c is a constant depending only on T.

By these lemmas and the expression (3) of ω, we can find a closed convex bounded subset S of $X_{\delta\alpha}$ such that $F^{(\nu)}S\subset S$ for small ν. On the other hand, $F^{(\nu)}$ is a compact operator in X_h for each $h>0$. Hence Schauder's theorem assures the existence of fixed points $\omega^{(\nu)}$ of $F^{(\nu)}$ in S. By lemma 2 we can also show that $\omega^{(\nu)}$ are uniformly bounded in X_α. The uniqueness follows from the next lemma.

Lemma 3. Suppose ω and ω' are fixed points of $F^{(\nu)}$ in X_α for some $\nu\geq 0$. Then for any $0\leq t<T$,

$$|F_1\omega(t,x) - F_1\omega'(t,x)| \leq Ct \sup_{0\leq s\leq t} \sup_{y\in R^2} |F_1\omega(s,y) - F_1\omega'(s,y)|$$

where C is a constant depending on ω_0, ω, ω', T and ν but not on t.

To go further to prove the convergence of the solutions $u^{(\nu)}$ of (NS) to $u^{(0)}$ of (E), we make the following assumptions on ω_0.

(3) $\quad F_1\omega_0 \in L^2(R^2)$

(4) $\quad$ as a function of x, $\sup_{|x-y|<1} \dfrac{|\omega_0(x)-\omega_0(y)|}{|x-y|^\alpha}$ is in $L^2(R^2)$.

Let $\omega^{(\nu)}$ be the fixed points of $F^{(\nu)}$, $\nu \geqq 0$, and $u^{(\nu)}=F_1\omega^{(\nu)}$. We obtain

Theorem 2. Additional to the assumption of theorem 1, assume ω_0 satisfies (3) and (4). Then for any β, $0<\beta<\alpha$, $u^{(\nu)} \to u^{(0)}$ in X_β as $\nu \to 0$.

Proof is done by showing L^2-convergence and using the interpolation of L^2 and B_α. We can also obtain the rate of convergence (see [4]).

Theorem 3. Under the assumptions of theorem 2, the following estimate is valid for any β, $0\leqq\beta<\alpha$.

$$\sup_{0\leqq t<T} |u^{(\nu)}(t,\cdot) - u^{(0)}(t,\cdot)|_{C^{1+h}(R^2)} = C\sqrt{\nu}^{\,\alpha-\beta}$$

where C is a constant depending on ω_0 and T.

Detailed proof of these theorems will be given elsewhere.

References

[1] K.K. Golovkin, Vanishing viscosity in Cauch's problem for hydromechanics equations, Proc. Steklov Inst. Math. 92 (1968) 33-53.

[2] T. Kato, Remarks on the Euler and Navier-Stokes equations in R^2.

[3] F.J. McGrath, Nonstationary plane flow of viscous and ideal fluids, Arch. Rational Mech. Anal. 27 (1968) 329-348.

[4] J.T.Beale and A. Majda, Rates of convergence for viscous Splitting of the Navier-Stokes equations, Math. Comp. 37 (1981) 243-259.

Lecture Notes in Num. Appl. Anal., 8, 133–161 (1985)
Recent Topics in Nonlinear PDE II, Sendai, 1985

Index Theorems and Bifucations in Duffing's Equations

Fumio Nakajima
Department of Mathematics, Iwate University
Morioka 020, Japan

1. Introduction.

We shall consider the Duffing's equation with periodic forcing term;

$$(0) \qquad \ddot{u} + k\dot{u} + au + bu^3 = B\cos\frac{2\pi}{\omega}t \qquad \left(\,\dot{} = \frac{d}{dt}\,\right)$$

where $k > 0$, $a \geqq 0$, $b > 0$, $B > 0$ and $\omega > 0$ are constants. It is known that equation (0) has several periodic solutions for appropriate k and B, which consist of harmonic solutions and subharmonic solutions, cf.[3],[4],[6],[9],[10]. The initial values at $t = 0$ of periodic solutions with period $n\omega$ are called $n\omega$-periodic points for integer $n > 0$ in the following, and they are fixed points of the n-th power of Poincaré mapping.

Several authors considered the indices of periodic points and obtained index theorems which show relationships among periodic points, cf.[2],[5],[7]. However their definition is done under the assumption that any periodic point is simple, that is, the modulus of characteristic multipliers of the periodic point are different from 1, and this assumption is not true in general for equation (0).

Recently G.Seifert and the author showed that the number of $n\omega$-periodic points of equation (0) is finite for each $n > 0$, cf.[8]. Moreover K.Shiraiwa at Nagoya University pointed out to the author that the above result enables us to define the indices of periodic points and state index theorems without the simplicity assumption. Theorem 1 and 2 are based on his idea. In Corollary 1, it is proved that if equation (0) has a directly unstable periodic solution, then there exist at least two other periodic solutions.

Index theorems seem to be useful to the study of bifucations of periodic solutions. In Theorem 3,4 and Corollary 3, we shall show that the bifucation of second order subharmonic solutions arises from the existence of an inversely unstable harmonic solution and that the bifucation of harmonic solutions arises from the existence of a direstly unstable harmonic solution. These results are illustrated by the physical data of [4] which will be stated later.

We denote by R^n the n-dimensional Euclidean space and set $R^1 = R$. By setting $\dot{u} = v$ in the equation (0), we have

$$(1) \quad \begin{cases} \dot{u} = v \\ \dot{v} = -kv - au - bu^3 + B\cos\frac{2\pi}{\omega}t \end{cases}$$

The system (1) is a particular case of (2)

$$(2)\quad \begin{cases} \dot{u} = U(t,u,v,B) \\ \dot{v} = V(t,u,v,B) \end{cases} \qquad (\cdot = \frac{d}{dt}),$$

where U(t,u,v,B) and V(t,u,v,B) are continuous for $(t,u,v,B) \in R^4$ and periodic in t with period $\omega > 0$;

$$\begin{aligned} U(t+\omega,u,v,B) &= U(t,u,v,B) \\ V(t+\omega,u,v,B) &= V(t,u,v,B) \end{aligned} \qquad \text{for } (t,u,v,B) \in R^4.$$

The parameter B is considered to be fixed in the section 2 and to be variable in the section 3.

Let (u(t,x,y),v(t,x,y)) be the solution of (2) through $(x,y) \in R^2$ at t = 0 for fixed B. (x,y) is an $n\omega$-periodic point if (u(t,x,y),v(t,x,y)) is periodic in t with period $n\omega$.

We assume that

(i) the system (2) is dissipative, that is, there is a compact subset D of R^2 such that any solution (u(t,x,y),v(t,x,y)) exists in the future and remains in D for large $t \geqq 0$,

(ii) U(t,u,v,B) and V(t,u,v,B) are analytic in u and v for fixed (t,B),

and

(iii) $\frac{\partial U}{\partial u} + \frac{\partial V}{\partial v} < 0$ for $(t,u,v,B) \in R^4$.

We can verify that (1) satisfies (i),(ii) and (iii).

2. Index theorems.

We shall consider Poincaré mapping $T : R^2 \to R^2$ defined by

$$T(x,y) = (u(\omega,x,y),v(\omega,x,y)).$$

it follows from Proposition that C has no ω-periodic point on it. Therefore I(T,C) is defined. If C is shrinked down to P continuously without crossing any ω-periodic point, the change of I(T,C) is also continuous. Since I(T,C) is an integer, it is constant. Therefore we can define

$$I(T,P) = I(T,C)$$

and call it the index of P by T. Similarly we can define $I(T^n,P)$ for $n\omega$-periodic point P.

Now we shall state relationships between the indices and the stabilities of ω-periodic point P = (x,y). Letting ρ_1 and ρ_2 be the characteristic multipliers of P, that is, the eigenvalues of the matrix

$$\begin{pmatrix} \frac{\partial u}{\partial x}(\omega,x,y) & \frac{\partial u}{\partial y}(\omega,x,y) \\ \frac{\partial v}{\partial x}(\omega,x,y) & \frac{\partial v}{\partial y}(\omega,x,y) \end{pmatrix},$$

we define that

(i) P is completely stable, if $|\rho_1| < 1$ and $|\rho_2| < 1$,

(ii) P is directly unstable if $0 < \rho_1 < 1 < \rho_2$

and

(iii) P is inversely unstable if $\rho_1 < -1 < \rho_2 < 0$.

The rest cases are the following;

(iv) $|\rho_1| > 1$ and $|\rho_2| > 1$

and

(v) $|\rho_1| = 1$ or $|\rho_2| = 1$.

The case (iv) should not happen, because condition (iii) implies $|\rho_1\rho_2| < 1$. On the other hand, the case (v) happens for system (1) where k,a,b and B are appropriate constants,

Since $u(t,x,y)$ and $v(t,x,y)$ are analytic in x and y by (ii), $T(x,y)$ is also analytic in x and y. Clearly,

$$T^n(x,y) = (u(n\omega,x,y),v(n\omega,x,y))$$

and (x,y) is an $n\omega$-periodic point if and only if

$$T^n(x,y) = (x,y).$$

We have obtained the following result in [8] ;

Proposition. For each integer $n > 0$, the number of $n\omega$-periodic points is finite.

Let C be a simply closed curve which has no ω-periodic point on it. For $Q_0 \in C$, let $Q_1 = TQ_0$ and $\overrightarrow{Q_0Q_1}$ be the vector from Q_0 to Q_1, which is not a zero vector. The number of revolutions made by $\overrightarrow{Q_0Q_1}$ as Q_0 traces out C once is an integer, since Q_0 returns to its starting position. Denoting the number by I_0, we set

$$I(T,C) = I_0$$

if the orientations of the revolutions made by Q_0 and by $\overrightarrow{Q_0Q_1}$ are the same, and

$$I(T,C) = -I_0$$

if they are reverse. $I(T,C)$ will be called the index of C by T. Similarly, the index of C by T^n, say $I(T^n,C)$, can be defined if C has no $n\omega$-periodic point on it. It should be noted that if the interior of C contains no $n\omega$-periodic point, then

$$I(T^n,C) = 0.$$

Now we shall consider ω-periodic point P. If C is a circle with center at P and with sufficiently small radius,

and it will be seen in the proof of Theorem 3 that this case is concerned with the bifucation of periodic solutions.
In [5], it is shown that if P is completely stable or inversely unstable, then $I(T,P) = +1$ and that if P is directly unstable then $I(T,P) = -1$. Since P is also a 2ω-periodic point, we may define $I(T^2,P)$. It should be noted that if P is inversely unstable, then

$$I(T^2,P) = -1,$$

which will be important in the proof of Theorem 3.

We shall state index theorems.

Theorem 1. Let n be any positive integer.

(i) If P is an $n\omega$-periodic point, then

$$|I(T^n,P)| \leqq 1.$$

(ii) If C is a simply closed curve which has no $n\omega$-periodic point on it and $\{P_j\}_{j=1}^{\alpha}$ is the set of $n\omega$-periodic points in the interior of C, then

$$I(T^n,C) = \sum_{j=1}^{\alpha} I(T^n,P_j).$$

(iii) If $\{P_j\}_{j=1}^{\beta}$ is the set of all $n\omega$-periodic points of (2), then

$$\sum_{j=1}^{\beta} I(T^n,P_j) = 1.$$

The proof is similar to one in [5] except assuming no simplicity condition.

The proof of (i). For the simplification we assume that $P = (0,0)$ and C is the circle with center at $(0,0)$ and with radius $r > 0$. The expansion of $T^n(u,v)$ around $(0,0)$ is that

$$\begin{pmatrix} u_1 \\ v_1 \end{pmatrix} = T^n \begin{pmatrix} u \\ v \end{pmatrix} = \begin{pmatrix} au + bv + o(u,v) \\ cu + dv + o(u,v) \end{pmatrix},$$

where a,b,c and d are constants such that

$$\begin{pmatrix} a & b \\ c & d \end{pmatrix} = \begin{pmatrix} \frac{\partial u}{\partial x}(n\omega, 0,0) & \frac{\partial u}{\partial y}(n\omega, 0,0) \\ \frac{\partial v}{\partial x}(n\omega, 0,0) & \frac{\partial v}{\partial y}(n\omega, 0,0) \end{pmatrix}$$

Since

$$0 < ad - bc < 1$$

by Abel's equality, we have at least following case 1 or case 2;

case 1 : $a \neq 1$ or $b \neq 0$

case 2 : $c \neq 0$ or $d \neq 1$.

Here we shall consider case 1, because the case 2 can be treated similarly. Denoting ϕ be the angle between the vector $(u_1 - u, v_1 - v)$ and u-axis, we have

$$\text{(3)} \qquad \phi = \tan^{-1} \frac{v_1 - v}{u_1 - u}$$

$$= \tan^{-1} \frac{(d-1)v + cu + o(u,v)}{(a-1)u + bv + o(u,v)} .$$

For $(u,v) \in C$, we set

$$u = r \cos \theta \qquad , v = r \sin \theta ,$$

where the domain of θ is $[0,2\pi)$. From (3) we have

$$(4) \qquad \phi = \tan^{-1}\frac{(d-1)\sin\theta + c\cos\theta + o(r)}{(a-1)\cos\theta + b\sin\theta + o(r)}.$$

Our purpose is to show that

$$(5) \qquad 0 \leqq \max_{0\leqq\theta\leqq 2\pi}\phi(\theta) - \min_{0\leqq\theta\leqq 2\pi}\phi(\theta) \leqq 2\pi.$$

Therefore, it is sufficient to show that for the denominator of the right hand side of (4), say

$$F(r,\theta) = (a-1)\cos\theta + b\sin\theta + o(r),$$

$F(r,\theta) = 0$ has exactly two roots in $[0,2\pi)$ for a small and fixed $r > 0$. We have

$$F(0,\theta) = (a-1)\cos\theta + b\sin\theta,$$

and hence $F(0,\theta) = 0$ has exactly two roots θ_1 and θ_2 in $[0,2\pi)$ such that

$$\theta_1 = \tan^{-1}\frac{1-a}{b}$$

and

$$\theta_2 = \theta_1 + \pi .$$

Since

$$\frac{\partial F}{\partial\theta}(0,\theta) = -(a-1)\sin\theta + b\cos\theta,$$

we have

$$\frac{\partial F}{\partial\theta}(0,\theta_i) \neq 0 \qquad \text{for } i = 1,2.$$

By the implicit function theorem, $F(r,\theta) = 0$ has exactly two roots $\theta_1(r)$ and $\theta_2(r)$ in the neighbourhoods of θ_1 and of θ_2, respectively, for sufficiently small $r > 0$, such that

$$\theta_1(0) = \theta_1 \quad \text{and} \quad \theta_2(0) = \theta_2.$$

Here, we understand that a neighbourhood of 0 is the union of neighbourhoods of 0 and 2π. This completes the proof of (i).

The proof of (ii). For each $n\omega$-periodic point P_j for $1 \leqq j \leqq \alpha$, we enclose it by a small circle C_j such that C_j is contained in the interior of C and

$$I(T^n, P_j) = I(T^n, C_j) .$$

Let us join C_j for $1 \leqq j \leqq \alpha$ and C together by curves so that if the curves are regarded as cuts, then the curves and cycles, C_j and C form a simple closed curve as illustrated in Fig.1 and denote this curve by Γ.

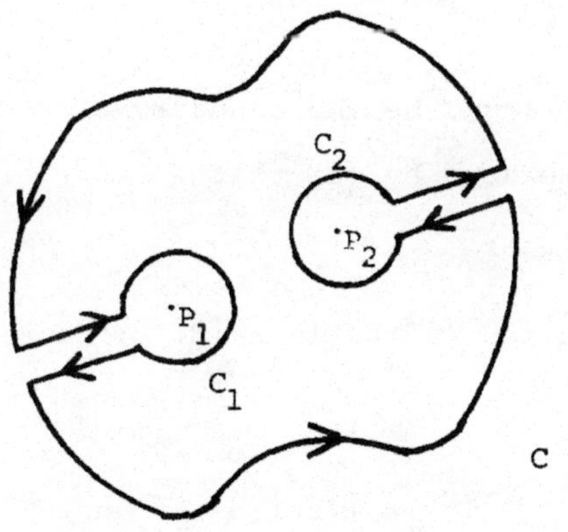

Fig.1

In determing the index of Γ, each curve segments, or cut, joining C_j and C is traversed first in one direction and then the other for $1 \leqq j \leqq \alpha$. Thus the net effect of these segments on the index is zero, and we have

$$I(T^n,\Gamma) = I(T^n,C) - \sum_{j=1}^{\alpha} I(T^n,C_j).$$

Since Γ contains no $n\omega$-periodic point in its interior, we have

$$I(T^n,\Gamma) = 0,$$

and hence

$$I(T^n,C) = \sum_{j=1}^{\alpha} I(T^n,C_j)$$

This completes the proof of (ii).

The proof of (iii). Since (2) is dissipative, there is a simply connected set K such that

$$(6) \qquad T^n K \subset K$$

as is stated in [5]. Letting C be the boundary of K which is a simply closed curve, we have by (6) that

$$I(T^n,C) = +1 .$$

Since $\{ P_j \}_{j=1}^{\beta}$ is contained in the interior of C, (ii) implies that

$$\sum_{j=1}^{\beta} I(T^n,P_j) = I(T^n,C) = +1 .$$

This completes the proof of (iii).

If P is an $n\omega$-periodic point, then $T^i P$ are also $n\omega$-periodic points for $1 \leqq i \leqq n-1$. Then we have following result.

Theorem 2.

(i) If P is an $n\omega$-periodic point, then

$$(7) \quad I(T^n, T^i P) = I(T^n, P) \quad \text{for } 1 \leqq i \leqq n-1$$

(ii) If $\{Q_j\}_{j=1}^{\alpha}$ is the set of $n\omega$-periodic points with least period $n\omega$, then

$$(8) \quad \sum_{j=1}^{\alpha} I(T^n, Q_j) \equiv 0 \quad (\text{ mod. } n)$$

Proof of (i). It is sufficient to show that for any $n\omega$-periodic point P,

$$I(T^n, P) = I(T^n, TP).$$

Let C_0 be a circle with center at P and with a sufficiently small radius such that it contains P as a unique $n\omega$-periodic point in its interior, and hence

$$(9) \quad I(T^n, P) = I(T^n, C_0).$$

Setting $C_t = \{ (u(t,Q), v(t,Q)) \in R^2 \; ; \; Q \in C_0 \}$ for $t \geqq 0$, by the uniqueness of solutions of initial value problems we can see that C_t is a simply closed curve and continuous for t. Since C_ω contains TP as a unique $n\omega$-periodic point in its interior, we have

$$(10) \quad I(T^n, C_\omega) = I(T^n, TP).$$

Now we shall consider the following mapping S_t on R^2 to R^2 ;

$$S_t(x,y) = (u(n\omega+t, t; x, y), \; v(n\omega+t, t; x, y)),$$

where (u(t,s;x,y),v(t,s;x,y)) is the solution of (2) through (x,y) at t = s. Clearly S_t is continuous for t and $S_0 = S_\omega = T^n$. For $Q \in C_t$, the solution through it at t is not nω-periodic, because it is on C_0 at t = 0, and hence

$$S_tQ \neq Q.$$

Therefore we can define the index of C_t by S_t as similar to I(T,C). In fact, when Q traces out C_t once, the number of revolution made by $\overrightarrow{Q\,Q_t}$, $Q_t = S_tQ$, is an integer, say I_0. We set

$$I(S_t,C_t) = I_0 \quad \text{and} \quad I(S_t,C_t) = -I_0 \quad ,$$

if the orientation of the revolutions $\overrightarrow{Q,Q_t}$ and Q are the same and if they are reverse, respectively. From the continuity of S_t and C_t for t, we can see that $I(S_t,C_t)$ is also continuous for t. Since $I(S_t,C_t)$ is an integer for each t, it is constant, and hence

$$I(S_\omega,C_\omega) = I(S_t,C_t) = I(S_0,C_0) \; .$$

By $S_\omega = S_0 = T^n$, we have

$$I(T^n,C_\omega) = I(T^n,C_0) \; .$$

Therefore (9) and (10) implies

$$I(T^n,TQ) = I(T^n,Q) \; .$$

This completes the proof of (i).

The proof of (ii). Since Q_j is a periodic point with least period $n\omega$, $\bigcup_{i=0}^{n-1}\{T^iQ_j\}$ is a set of distinct $n\omega$-periodic points. Therefore $\{Q_j\}_{j=1}^{\alpha}$ can be rearranged as a disjoint union;

$$\{Q_j\}_{j=1}^{\alpha} = \bigcup_{k=1}^{\beta}\bigcup_{i=0}^{n-1}\{T^iQ_{j_k}\},$$

where $\{Q_{j_k}\}_{k=1}^{\beta}$ is a suitable subset of $\{Q_j\}_{j=1}^{\alpha}$.

We have

$$\sum_{j=1}^{\alpha} I(T^n,Q_j) = \sum_{k=1}^{\beta}\{\sum_{i=0}^{n-1} I(T^n,T^iQ_{j_k})\}.$$

Since $I(T^n,T^iQ_{j_k}) = I(T^n,Q_{j_k})$ for $0 \leqq i \leqq n-1$ by the above (i), we have

$$\sum_{i=1}^{\alpha} I(T^n,Q_j) = \sum_{k=1}^{\beta} nI(T^n,Q_{j_k})$$

$$= n\sum_{k=1}^{\beta} I(T^n,Q_{j_k})$$

which shows (8).

We have following corollaries.

<u>Corollary 1</u>. If (2) has a directly unstable ω-periodic point, then there exist at least two other ω-periodic points.

Proof. Let P_1 be the directly unstable ω-periodic point. Since $I(T,P_1) = -1$, (iii) of Theorem 1 implies

$$\sum_{j=2}^{\beta} I(T,P_j) = 1 - I(T,P_1) = 2.$$

Therefore, (i) of Theorem 1 implies the existence of two ω-periodic points with index +1.

The above result is illustrated in Fig.2 by the data given in [4] which shows the location of 2π-periodic points of the equation
$\ddot{u} + 0.2\dot{u} + u^3 = 0.3\cos t$.

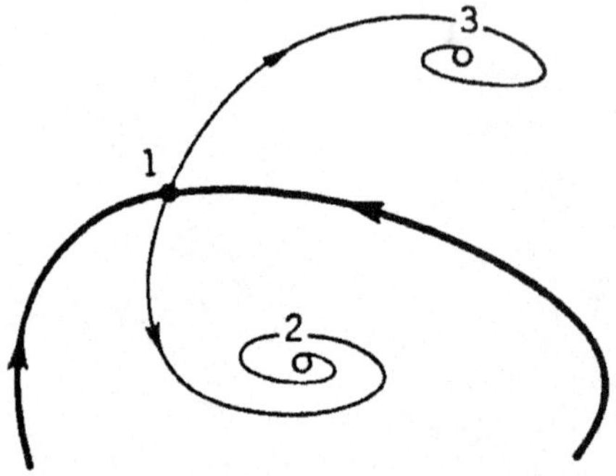

Figure 2.

Here point 1 is a directly unstable 2π-periodic point, and points 2 and 3 are 2π-periodic points with index +1.

Corollary 2. Let α be the number of ω-periodic points of (2). If n is a prime number such that

$$n > \alpha + 1,$$

then for the set of periodic points with least period $n\omega$, say $\{Q_k\}_{k=1}^{\beta}$, we have

$$\sum_{k=1}^{\beta} I(T^n, Q_k) = 0. \tag{11}$$

Consequently, all the periodic points of least period of $n\omega$ cannot be completely stable.

Proof. We denote the set of ω-periodic points by $\{P_j\}_{j=1}^{\alpha}$. Since n is a prime number, the set of $n\omega$-periodic points consists of only $\{P_j\}_{j=1}^{\alpha}$ and $\{Q_k\}_{k=1}^{\beta}$. By (iii) of Theorem 1, we have

$$\sum_{j=1}^{\alpha} I(T^n, P_j) + \sum_{k=1}^{\beta} I(T^n, Q_k) = +1 ,$$

and hence

$$\sum_{k=1}^{\beta} I(T^n, Q_k) = 1 - \sum_{j=1}^{\alpha} I(T^n, P_j) .$$

Since $|I(T^n, P_j)| \leqq 1$, we have

$$\left| \sum_{k=1}^{\beta} I(T^n, Q_k) \right| \leqq 1 + \alpha .$$

By (ii) of Theorem 2, there is an integer q such that

$$\sum_{k=1}^{\beta} I(T^n, Q_k) = nq .$$

Therefore we have

$$|nq| \leqq 1 + \alpha .$$

Since $n > \alpha + 1$, we obtain $q = 0$ which shows (11).

Remark. In [7], (11) is proved for all odd number n under the assumption that every periodic point is simple.

3. Bifucations.

A second order subharmonic point is the initial value of second order subharmonic solution at $t = 0$, that is, it is a 2ω-periodic point but not a ω-periodic point. In the following, the parameter B of (2) is variable and the Poincaré mapping and ω-periodic point are denoted by $T(B)$ and $P(B)$, respectively. Since $U(t,x,y,B)$ and $V(t,x,y,B)$ are continuous for $(t,x,y,B) \in R^4$, $T(B)(x,y)$ is continuous in $(x,y,B) \in R^3$.

<u>Theorem 3</u>. We assume that there is a $B_0 \in R$ and $\varepsilon_0 > 0$ such that (2) has an ω-periodic point $P(B)$ which is continuous in $B_0 - \varepsilon_0 < B < B_0 + \varepsilon_0$, completely stable for $B_0 - \varepsilon_0 < B < B_0$ and inversely unstable for $B_0 < B < B_0 + \varepsilon_0$.

Then , for a sufficiently small $\varepsilon > 0$, we have following conclusions (i), (ii) or both ;

(i) for any $B, B_0 < B < B_0 + \varepsilon$, (2) has at least two second order subharmonic points $Q_1(B)$ and $Q_2(B)$ such that

$$Q_i(B) \longrightarrow P(B_0) \qquad \text{as } B \longrightarrow B_0$$

and

$$I(T^2(B), Q_i(B)) = -I(T^2(B), P(B))$$

for $i = 1$ and $i = 2$,

(ii) for any B, $B_0 - \varepsilon < B < B_0$, the same as in (i) holds.

The above result may be illustrated by the following figures.

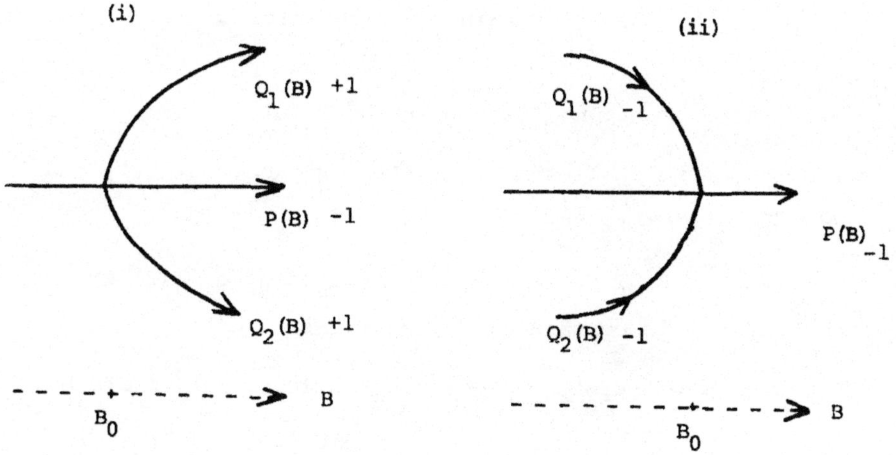

Fig.3

For example, in (i), the arrows ⟶ represent the manners how $Q_1(B)$ and $Q_2(B)$ bifucate from P(B) as B is increasing. The number +1 or -1 attached to $Q_1(B)$, $Q_2(B)$, P(B) denotes the index of the point by $T^2(B)$.

We shall prove Theorem 3 by three steps.

Step 1. Letting $P_0 = P(B_0)$, we can see that P_0 is a fixed point of $T(B_0)$ and $T^2(B_0)$. By Proposition, there is a circle C with center at P_0 and with a small radius such that C has no fixed point of $T^2(B_0)$ on it and P_0 is a unique fixed point of $T^2(B_0)$ in the interior of C.

Therefore we have

$$(12) \qquad I(T^2(B_0), P_0) = I(T^2(B_0), C) .$$

If ε is sufficiently small, then it follows from the continuity of P(B) and T(B) that if $|B - B_0| < \varepsilon$,

(*) P(B) is contained in the interior of C

and

(**) C has no fixed point of $T^2(B)$ on it.

Now we shall show that if ε is sufficiently small and $|B - B_0| < \varepsilon$, then

(***) P(B) is a unique fixed point of T(B) which is contained in the interior of C.

In fact, fixed points of T(B) are the zero points of the following function;

$$S(x,y,B) = T(B)(x,y) - (x,y) .$$

We have

$$\frac{\partial S(x,y,B_0)}{\partial(x,y)} = (\rho_1 - 1)(\rho_2 - 1) \quad \text{at } (x,y) = P_0$$

where ρ_1 and ρ_2 are the characteristic multipliers of P_0.

Since P(B) is inversely unstable for $B_0 < B < B_0 + \varepsilon$, we have that

$$\rho_1 \leqq -1 \leqq \rho_2 < 0 \quad ,$$

which implies

$$\frac{\partial S(x,y,B_0)}{\partial(x,y)} \neq 0 \qquad \text{at } (x,y) = P_0.$$

By the implicit function theorem, the zero point of S(x,y,B) is unique in the neighbourhood of P_0 for $|B - B_0| < \varepsilon$ if ε is sufficiently small. Therefore, taking the radius of C so small that its interior is contained in the above neighbourhood of P_0, we have (***) .

Step 2. In the following, we take ε so small that (*),(**) and (***) hold, and let $|B - B_0| < \varepsilon$. By (**), $I(T^2(B),C)$ is defined and continuous for B. Since $I(T^2(B),C)$ is an integer, it is constant, and hence

$$(13) \quad I(T^2(B),C) = I(T^2(B_0),C) .$$

Letting $\{P_j\}_{j=1}^{\alpha}$ be the set of fixed points of $T^2(B)$ contained in the interior of C, we have from (ii) of Theorem 1 that

$$(14) \quad I(T^2(B),C) = \sum_{j=1}^{\alpha} I(T^2(B),P_j) ,$$

and it follows from (12),(13) and (14) that

$$(15) \quad I(T^2(B_0),P_0) = \sum_{j=1}^{\alpha} I(T^2(B),P_j) .$$

By (*), P(B) is contained in the interior of C , and hence

$$P(B) = P_j \quad \text{for some } j.$$

Since P(B) is completely stable for $B_0 - \varepsilon < B < B_0$ and inversely unstable for $B_0 < B < B_0 + \varepsilon$, we have

$$I(T^2(B),P(B)) = \begin{cases} +1 & \text{for } B_0 - \varepsilon < B < B_0 , \\ -1 & \text{for } B_0 < B < B_0 + \varepsilon. \end{cases}$$

We shall show that if $B_0 - \varepsilon < B < B_0$, $B_0 < B < B_0 + \varepsilon$ or $B \neq B_0$, then $T^2(B)$ has at least two fixed points in the interior of C whose indices are +1 and -1. For $B_0 - \varepsilon < B < B_0$, suppose that $T^2(B)$ has no fixed point with index -1. Then we have by (i) of Theorem 1 that

$$I(T^2(B),P_j) \geqq 0 \qquad \text{for } 1 \leqq j \leqq \alpha,$$

and hence , by (15)

$$I(T^2(B_0),P_0) \geqq I(T^2(B),P(B)) = +1$$

which shows by (i) of Theorem 1 that

$$(16) \qquad I(T^2(B_0),P_0) = +1 .$$

Similarly, supposing that $T^2(B)$ has no fixed point with index +1 in the interior of C for some B, $B_0 < B < B_0 + \varepsilon$, we can show by the same argument as above that

$$I(T^2(B_0),P_0) = -1$$

which contradicts to (16). Therefore we obtain the above assertion.

Since the fixed point of T(B) is unique in the interior of C by (***), one of the above two fixed pints of $T^2(B)$ is not a fixed point of T(B) , and therefore it is a second order subharmonic point.

Step 3. Letting the second order subharmonic point be $Q_1(B)$ and $Q_2(B) = T(B)Q_1(B)$, we have by (i) of Theorem 2 that $Q_2(B)$ is a second order subharmonic point and

$$I(T^2(B),Q_1(B)) = I(T^2(B),Q_2(B)).$$

In the above argument, we have seen that

$$I(T^2(B),Q_1(B)) = -I(T^2(B),P(B)),$$

and therefore

$$I(T^2(B), Q_i(B)) = -I(T^2(B), P(B)) \quad \text{for } i = 1 \text{ and } i = 2,$$

Now, letting $B \longrightarrow B_0$, we have

$$P(B) \longrightarrow P_0 \ .$$

In the above arguments, the radius of C is only required to satisfy that C contains P(B) in its interior. Since $Q_1(B)$ is also contained in the interior of C, we have

$$Q_1(B) \longrightarrow P_0 \quad \text{as } B \longrightarrow B_0 \ ,$$

and hence

$$\lim_{B \to B_0} Q_2(B) = \lim_{B \to B_0} T(B) Q_1(B) = T(B_0) Q_1(B_0) =$$

$$= T(B_0) P_0 = P_0 \ .$$

This completes the proof.

Replacing $T^2(B)$ by T(B) in the above proof, we can obtain the following result.

Corollary 3. For some $B_0 \in R$ and $\varepsilon_0 > 0$, we assume that (2) has an ω-periodic point P(B) which is continuous for $B_0 - \varepsilon_0 < B < B_0 + \varepsilon_0$, completly stable for $B_0 - \varepsilon_0 < B < B_0$ and directly unstable for $B_0 < B < B_0 + \varepsilon_0$.

Then for sufficiently small $\varepsilon > 0$, we have following conclusions (i), (ii) or both ;

(i) for $B_0 < B < B_0 + \varepsilon$, (2) has at least one ω-periodic point $Q(B)$ such that

$$Q(B) \neq P(B),$$

$$I(T(B), Q(B)) = -I(T(B), P(B))$$

and

$$Q(B) \longrightarrow P(B_0) \qquad \text{as } B \longrightarrow B_0 ,$$

(ii) for $B_0 - \varepsilon < B < B_0$, the same as in (i) holds.

We shall return to the Duffing's equation ;

$$(1) \quad \begin{cases} \dot{u} = v \\ \dot{v} = -kv - au - bu^3 + B\cos\frac{2\pi}{\omega}t . \end{cases}$$

Theorem 4. If there is a parameter B^* such that (1) has inversely unstable ω-periodic point for $B = B^*$, then there is a B_0, $0 < B_0 < B^*$, and a small number $\varepsilon_0 > 0$ such that (1) has a periodic point $P(B)$ which is analytic for $B_0 - \varepsilon_0 < B < B_0 + \varepsilon_0$, completely stable for $B_0 - \varepsilon_0 < B < B_0$ and inversely unstable for $B_0 < B < B_0 + \varepsilon_0$. Consequently the conclusion of Theorem 3 holds.

Remark. In the above we considered the case where B is increasing for a fixed k. The same conclusion can be obtained by considering that k is increasing for a fixed B under the same assumptions as of Theorem 4.

We shall prove the theorem by three steps.

Step 1. Let $(u(t,x,y,B), v(t,x,y,B))$ be the solution of (1) through (x,y) at $t = 0$ and set

$$f(x,y,B) = u(\omega,x,y,B) - x$$

$$g(x,y,B) = v(\omega,x,y,B) - y .$$

Since $u(t,x,y,B)$ and $v(t,x,y,B)$ are anlytic for $(x,y,B) \in R^3$, $f(x,y,B)$ and $g(x,y,B)$ are analytic for (x,y,B). Clearly (x,y) is an ω-periodic point if and only if

$$f(x,y,B) = g(x,y,B) = 0.$$

Let (x^*,y^*) be the given inversely unstable ω-periodic point for $B = B^*$ and ρ_1, ρ_2 be the characteristic multipliers such that

$$\rho_1 < -1 < \rho_2 < 0 .$$

We have

$$f(x^*,y^*,B^*) = g(x^*,y^*,B^*) = 0$$

and

$$\frac{\partial(f,g)}{\partial(x,y)}(x^*,y^*,B^*) = (\rho_1 - 1)(\rho_2 - 1) \neq 0.$$

Therefore, by the implicit function theorem, there exists analytic functions $x(B)$ and $y(B)$ defined in the neighbourhood of B^* such that

$$(17)\quad \begin{cases} f(x(B),y(B),B) = 0 \\ g(x(B),y(B),B) = 0 \end{cases}$$

and

$$X(B^*) = x^* \;,\; y(B^*) = y^* \;.$$

Considering the analytic continuation of $(x(B),y(B))$ for $B < B^*$, we have the maximal interval (B_1,B^*) such that $(x(B),y(B))$ is analytic for $B_1 < B < B^*$. Since (17) is satisfied in the neighbourhood of B^*, it follows from the unicity theorem that (17) holds on (B_1,B^*), and hence $(x(B),y(B))$ is an ω-periodic point on (B_1,B^*).

Step 2. Let $\rho_1(B)$ and $\rho_2(B)$ be the characteristic multipliers of $(x(B),y(B))$ such that they are ordered to be continuous for B and

$$\rho_1(B^*) < -1 < \rho_2(B^*) < 0.$$

By Abel's eqality, we have

$$(18)\qquad \rho_1(B)\rho_2(B) = e^{-\omega k} \;.$$

We shall show that there is a $B_2 \in (B_1,B^*)$ such that

$$(19)\qquad -1 < \rho_1(B_2) < 0 \;.$$

To the contrary we suppose that

$$(20)\qquad \rho_1(B) \leqq -1 \quad \text{for all } B \in (B_1,B^*).$$

If $B_1 > -\infty$, $P(B)$ is bounded on (B_1,B^*) and has accumulation points as $B \longrightarrow B_1$. Letting (x_1,y_1) be one of them, we have

$$f(x_1, y_1, B_1) = 0$$

and

$$g(x_1, y_1, B_1) = 0.$$

Since the characteristic multipliers ρ_1 and ρ_2 of (x_1, y_1) satisfy by (18) and (20)

$$\rho_1 \leqq -1 \leqq \rho_2 < 0 ,$$

it follows from the same argument as of Step 1 that P(B) can be defined to be analytic for $B_1 - \varepsilon_1 < B < B_1$ and for a small number $\varepsilon_1 > 0$. This contradicts to the definition of B_1. Therefore we have that

$$B_1 = -\infty ,$$

and hence

$$0 \in (-\infty, B^*) = (B_1, B^*)$$

which implies

$$\rho_1(0) \leqq -1 \leqq \rho_2(0) < 0. \tag{21}$$

On the other hand, for B = 0, (1) is reduced to the system ;

$$\begin{cases} \dot{u} = v \\ \dot{v} = -kv - au - bu^3 . \end{cases}$$

Clearly, this system has the unique periodic point (0,0). Therefore $\rho_1(0)$ and $\rho_2(0)$ is the characteristic multipliers of (0,0) which is completely stable, and hence

$$|\rho_1(0)| < 1 \quad \text{and} \quad |\rho_2(0)| < 1$$

which contradicts to (21). Thus (19) is proved.

Step 3. Since $\rho_1(B)$ is continuous for B,by (19) there is a $B \in (B_2, B^*)$ such that

$$(22) \qquad \rho_1(B) = -1$$

which implies with (18)

$$\rho_2(B) = -e^{-\omega k} \quad .$$

Since $\rho_1(B) \neq \rho_2(B)$, it is known that $\rho_1(B)$ and $\rho_2(B)$ are analytic at B. Therefore there is a $B_0 \in (B_2, B^*)$ and a small number $\varepsilon > 0$ such that

$$-1 < \rho_1(B) < 0 \qquad \text{for } B_0 - \varepsilon < B < B_0 ,$$

$$\rho_1(B_0) = -1$$

and

$$\rho_1(B) < -1 \qquad \text{for } B_0 < B < B_0 + \varepsilon .$$

Therefore P(B) is completely stable for $B_0 - \varepsilon < B < B_0$ and inversely unstable for $B_0 < B < B_0 + \varepsilon$. This complete the proof.

Thoerem 4 is illustrated by the following data of [4] which shows the location of 2π periodic points and second order subharmonic point of the equation

$$\ddot{u} + 0.2\dot{u} + u^3 = B \cos t ,$$

where B is increasing.

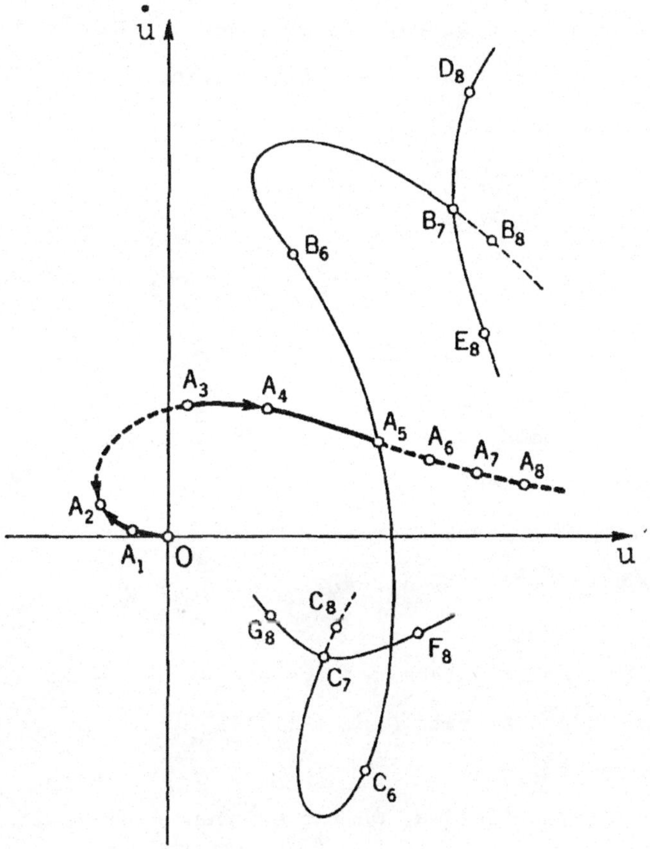

When B = 0.3, there is a completely stable 2π-periodic point B_6. When B = 3, there is a 2π-periodic point B_7 from which two second order subharmonic points bifucate. When B = 5.5 , there are an inversely unstable 2π-periodic point B_8 and two completely satble second order subharmonic points D_8 and E_8.

Acknowledgement. The author wishes his invaluable thanks to Professor K.Shiraiwa at Nagoya University for his comments and suggestions. Moreover the author wishes his invaluable thanks to Professors C.Hayashi, Y.Ueda and H.Kawakami for thier permitting that thier interesting data of [4] may be used here.

References

[1] K.T.Alligood, J.Mallet-Paret and J.A.York, An index for the global continuation of relatively isolated sets of periodic orbits, Geometric Dynamics, Springer Lecture Note in Math. 1007(1983)

[2] G.D.Birkhoff, Dynamical systems with two degree of freedom, Trans.Am.Math.Soc. 18(1917)

[3] Funato and Maekawa, On the existence of subharmonics for Duffing's equation, Math.Japonica, 5(1958 ∼ 59), pp.27-32.

[4] C.Hayashi, Y.Ueda and H.Kawakami, Transformation theory as applied to the solutions of nonlinear differential equations of the second order, Int. J.Non-linear Mechanics, vol.4(1969),pp.235-255.

[5] N.Levinson, Transformation theory of nonlinear differtial equations of the second order, Ann. Math., 45(1944), pp.723 - 737.

[6] W.S.Loud, Periodic solutions of $\ddot{x} + c\dot{x} + g(x) = \varepsilon f(t)$, Amer. Math. Soc. Mem., No. 31(1958).

[7] J.L.Massera, The number of subharmonic solutions of nonlinear differential equations of the second order, Ann. Math., 50(1949), pp. 118 - 126.

[8] F.Nakajima and G.Seifert, On the number of periodic solutions of 2-dimensional periodic sytems, J. Diff. Equations, vol. 49, No.3(1983), pp.430 - 440.

[9] Y.Shinohara, Numerical investigation of $\frac{1}{2}$ - subharmonic solutions to Duffing's equation, Memoirs of Numerical Mathematics, No. 1(1974).

[10] M.Urabe, Numerical investigation of subharmonic solutions to Duffing's equation, Publ. RIMS, Kyoto Univ., vol.5 (1969), pp.79 - 112.

This Page Intentionally Left Blank

Lecture Notes in Num. Appl. Anal., 8, 163–174 (1985)
Recent Topics in Nonlinear PDE II, Sendai, 1985

Global Solutions for Some Nonlinear Parabolic Equations with Non-monotonic Perturbation

Mitsuhiro NAKAO

Department of Mathematics, College of General Education, Kyushu University, Fukuoka 810, Japan

0. Introduction

In this article we are concerned with the existence, uniqueness and decay property of the solutions of the following two problems

$$(P_1)\quad \begin{cases} u_t - \sum_{i=1}^{N} (|u_{x_i}|^m u_{x_i})_{x_i} + \beta(x,u) = 0 \quad \text{in } \Omega \times R^+ & (1) \\ u(x,0) = u_0 \quad \text{and} \quad u(x,t)|_{\partial\Omega} = 0 \end{cases}$$

and

$$(P_2)\quad \begin{cases} u_t - \Delta(|u|^m u) + \beta(x,u) = 0 \quad \text{in } \Omega \times R^+ & (2) \\ u(x,0) = u_0 \quad \text{and} \quad u(x,t)|_{\partial\Omega} = 0 \end{cases}$$

where Ω is a bounded domain in R^n with smooth boundary $\partial\Omega$ and $\beta(x,u)$ is a (locally) Hölder continuous function on $\bar{\Omega} \times R^+$ such that

$$|\beta(x,u)| \leq k_0 |u|^{\alpha+1} \qquad (3)$$

with some $\alpha \geq 0$ and $k_0 > 0$.

The existence and the nonexistence of global solutions for (P_1) with $\beta(x,u)=-|u|^{\alpha}u$ were first investigated by Tsutsumi [11] using the concept of the ' potential well' introduced by Sattinger [9]. The essential assumption in [11] for global existence is the growth condition on $\beta(x,u)$, i.e.,

$$0 \leq \alpha < m_0^* \tag{4}$$

where

$$m_0^* = \begin{cases} (mN+2m+4)/(N-m-2) & \text{if } N > m+2 \\ \text{arbitrarily large} & \text{if } 1 \leq N \leq m+2. \end{cases} \tag{5}$$

The result of [11] was generalized by Ôtani [8], Ishii [3] and Nakao & Narazaki [7] etc.. Concerning the uniqueness very little is known, i.e., the solution is known to be unique only for the case $1 \leq N < m+2$. (This case is rather trivial by the inclusion $W_0^{1,m+2} \subset L^{\infty}$).

A parallel result to [11] was proved also for the problem (P_2) by Galaktinov [2] and Nakao [4] independently. For (P_2) the restriction on α is

$$0 \leq \alpha < m_1^* \tag{6}$$

where

$$m_1^* = \begin{cases} (mN+4)/(N-2) & \text{if } N > 2 \\ \text{arbitrarily large} & \text{if } N=1,2. \end{cases} \tag{7}$$

The decay properties of solutions of (P_1) and (P_2) were discussed in [7] and [4], respectively (see also [3]).

Recently in [5] we have derived L^∞ estimate of the solutions both for (P_1) and (P_2) using the so-called Moser's technique, and consequently proved the existence and uniqueness theorem for (P_1) and (P_2) under the assumption (4) and (6), respectively. Saks [10] also has proved closely related result for (P_2) with the use of quitely different method. In [10] the assumption (6) is made implicitely.

The object of this article is to show the existence, uniqueness and some decay properties of the solutions of (P_1) and (P_2) without any restriction on α.

1. Some lemmas and the statement of result.

First we recall the Sobolev's Lemma and (a variant of) Gagliardo-Nirenberg inequality.

Lemma 1. $W_0^{1,p}(\Omega)$ is continuously embedded into $L^q(\Omega)$ provided that (i) $N > p \geq 1$ and $1 \leq q \leq Np/(N-p)$ (ii) $N=p>1$ and $1 \leq q < \infty$ or (iii) $1 \leq N < p$ and $1 \leq q \leq \infty$.

Lemma 2. For all u with $|u|^\gamma u \in W^{1,p}(\Omega)$, $\gamma \geq 0, p \geq 1$, we have

$$\|u\|_q \leq C\|u\|_r^{1-\theta}\|\nabla(|u|^\gamma u)\|_p^{\theta/(\gamma+1)}$$

with $C > 0$ and $\theta = (\gamma+1)(r^{-1}-q^{-1})\{N^{-1}-p^{-1}+(\gamma+1)r^{-1}\}^{-1}$ provided that $q \geq \gamma+1$ and (i) $1 \leq r \leq q \leq (\gamma+1)Np/(N-p)$ if $N > p$ (ii) $1 \leq r \leq q < \infty$ if $N=p=1$ or (iii) $1 \leq r \leq q \leq \infty$ if $1 \leq N < p$.
(The case $\gamma=0$ is the original G-N inequality.)

For the L^∞ - boundedness of solutions the following lemma will play an essential role.

Lemma 3. Let $w(t)\equiv w(x,t)$ be an appropriately smooth function defined on $\bar{\Omega}\times R^+$, satisfying

$$\frac{d}{dt}\|w(t)\|_{\lambda+1}^{\lambda+1} + C_o(1+\lambda)^{-\theta_o}\|\nabla(|w(t)|^{(\lambda+m+1-r)/r}w(t))\|_r^r$$

$$\leq C_1(1+\lambda)^{\theta_1}\|w(t)\|_{\lambda+1}^{\lambda+1}$$

for any $\lambda \geq \lambda_o > \max(0, r-m-1, (m-r+1)/(r-1))$ with some constants $C_o(>0), C_1(>0)$, $\theta_o(\geq 0)$, $\theta_1(\geq 0)$ and $r \geq 1$. Suppose that $w_o \equiv w(0) \in L^\infty(\Omega)$ and $\sup_{t\geq 0}\|w(t)\|_{\lambda_o+1} < \infty$. Then, there exists $C>0$ and $d>0$ such that

$$\sup_{t\geq 0}\|w(t)\|_\infty \leq C\max(1, \sup_{t\geq 0}\|w(t)\|_{\lambda_o+1}^d, \|w_o\|_\infty).$$

Lemma 3 is a generalization of Alikakos [1;Lemma 3.2] and can be provrd by Moser's technique (cf.[5 ; Appendix]).

Our result reads as follows.

Theorem 1. Let $\alpha > m$ and let β satisfy (3). Assume that $u_o \in W_o^{1,m+2}\cap L^{p_o+2}$ for some p_o with $p_o \geq 0$ and $p_o > \frac{N(\alpha-m)}{m+2}-2$. Then, there exists $d_o>0$ such that if $\|u_o\|_{p_o+2} < d_o$ the problem (P_1) admits a global solution $u(t)$ which satisfies

$$u\in L^\infty(R^+; W_o^{1,m+2}\cap L^{p_o+2}),\quad u_t\in L^2(R^+;L^2)$$

and

$$\|\nabla u(t)\|_{m+2} + \|u(t)\|_{p_o+2} + \left(\int_t^\infty \|u_t(s)\|_2^2 ds\right)^{1/(m+2)} \leq C(1+t)^{-1/m} \quad (8)$$

with some C depending on $\|\nabla u_o\|_{m+2}$.

Moreover, under the additional assumption $u_o \in L^{\infty}(\Omega)$, we have $u \in L^{\infty}(R^+; W_o^{1,m+2} \cap L^{\infty})$ and

$$\|u(t)\|_{\infty} \leq C(1+t)^{-1/m}. \tag{9}$$

Such solution is unique if $\beta(x,u)$ is locally Lipshitzian in u.

Theorem 2. Let $\alpha > m$ and β satisfy (3). Assumethat $|u_o|^m u_o \in \mathring{H}_1 \cap L^{(p_1+2)(m+1)}$ with p_1 such that $p_1 \geq 0$ and $p_1 > \frac{N(\alpha-m)}{2} - 2$. Then, there exists $d_1 > 0$ such that if $\|u_o\|_{p_1+2} < d_1$ the problem (P_2) admits a solution u(t) satisfying;

$$|u|^m u \in L^{\infty}(R^+; \mathring{H}_1 \cap L^{(p_1+2)(m+1)}), \quad \frac{\partial}{\partial t}(|u|^{m/2}u) \in L^2(R^+; L^2)$$

and

$$\|\nabla(|u|^m u(t))\|_2 + \|u(t)\|_{p_1+2}^{m+1} + \int_t^{\infty} \|\frac{\partial}{\partial t}(|u|^{m/2}u)\|_2^2 ds$$

$$\leq C(1+t)^{-(m+1)/m}. \tag{10}$$

Moreover, under the additional assumption $u_o \in L^{\infty}$ the solution u belongs to $L^{\infty}(R^+; L^{\infty})$ and satisfies

$$\|u(t)\|_{\infty} \leq C(1+t)^{-1/m}. \tag{11}$$

Such solution is unique if $\beta(x,u)$ is locally Lipshitzian in u.

2. Outline of the proof of Theorem 1.

The solution will be given as a limit of smooth approximate solutions. For this we must derive the estimates (8) and (9) for the approximate solutions. Here we shall give an outline of the proof of such estimates for (assumed) smooth solution (P_1). We write u^q for $|u|^{q-1}u, q \geq 1$.

Multiplying the equation (1) by $u^{p+1}, p \geq 0$, we have

$$\frac{1}{p+2}\frac{d}{dt}\|u(t)\|_{p+2}^{p+2} + \frac{(p+1)(m+2)^{m+2}}{(p+m+2)^{m+2}}\|\nabla u^{(p+m+2)/(m+2)}\|_{m+2}^{m+2}$$

$$\leq k_0\|u\|_{p+\alpha+2}^{p+\alpha+2} . \tag{12}$$

Here we utilize Lemmas 1 and 2 to get

$$\|u\|_{p_0+\alpha+2}^{p_0+\alpha+2} \leq C_0\|u\|_{p_0+2}^{\alpha-m}\|\nabla u^{(p_0+m+2)/(m+2)}\|_{m+2}^{m+2} \tag{13}$$

for some $C_0 > 0$ under the assumption on p_0. From (12) and (13) we obtain

$$\frac{d}{dt}\|u(t)\|_{p_0+2}^{p_0+2} \leq C_0(p_0+2)\left\{-(d_0')^{\alpha-m} + \|u(t)\|_{p_0+2}^{\alpha-m}\right\}\|\nabla u^{(p_0+m+2)/(m+2)}\|_{m+2}^{m+2} \tag{14}$$

where we set

$$d_0' \equiv \left(\frac{(m+2)^{m+2}(p_0+1)}{C_0(p_0+m+2)^{m+2}}\right)^{1/(\alpha-m)} .$$

Now, making the assumption $\|u_o\|_{p_o+2} < d_o'$, we conclude that $\|u(t)\|_{p_o+2}$ is monotone decreasing and

$$\frac{d}{dt}\|u(t)\|_{p_o+2}^{p_o+2} + \varepsilon_o \|\nabla u^{(p_o+m+2)/(m+2)}\|_{m+2}^{m+2} \leq 0 \tag{15}$$

for a certain $\varepsilon_o > 0$. This inequality implies , in particular,

$$\|u(t)\|_{p_o+2} \leq C(1+t)^{-1/m} . \tag{16}$$

Next, multiplying the equation (1) by u_t we have

$$\frac{d}{dt} J_o(u(t)) + \|u_t(t)\|_2^2 = 0 \tag{17}$$

where we set

$$J_o(u(t)) = \frac{1}{m+2}\|\nabla u(t)\|_{m+2}^{m+2} + \int_\Omega \int_0^u \beta(x,\eta)\,d\eta\,dx.$$

Here, note that if $N > m+2$ and $\alpha \geq m_o^*$ (and consequently $p_o \geq \alpha$) we see $\|u(t)\|_{\alpha+2} \leq Cd_o' < \infty$. When $p_o < \alpha$ we have, by Lemmas 1 and 2, that

$$\|u\|_{\alpha+2}^{\alpha+2} \leq C_1 \|u\|_{p_o+2}^{\alpha-m} \|\nabla u\|_{m+2}^{m+2} \tag{18}$$

for a certain $C_1 > 0$. Hence, under the additional assumption

$$\|u_o\|_{p_o+2} < d_o'' \equiv \left(\frac{\alpha+2}{C_1(m+2)}\right)^{1/(\alpha-m)} \tag{19}$$

we have

$$J_o(u(t)) \geq \varepsilon_1 \|\nabla u(t)\|_{m+2}^{m+2} \tag{20}$$

for a certain $\varepsilon_1 > 0$.

Thus, setting $d_o = d_o'$ if $p_o \geq \alpha$ and $d_o = \min(d_o', d_o'')$ if $p_o < \alpha$ we obtain the boundedness of $\|\nabla u(t)\|_{m+2}$.

Since

$$\|u(t)\|_{\alpha+2}^{\alpha+2} \leq C\|u(t)\|_{p_o+2}^{\alpha-m} \|\nabla u(t)\|_{m+2}^{m+2} \leq C(1+t)^{-(\alpha-m)/m} \|\nabla u\|_{m+2}^{m+2} \tag{21}$$

we can take $T_o > 0$ such that (20) is valid for $t \geq T_o$ and also for $p_o \geq \alpha$. Next, multiplying the equation by u and using (21) we may assume

$$\int_\Omega uu_t dx + \varepsilon_2 \|\nabla u(t)\|_{m+2}^{m+2} \leq 0 \text{ for } t \geq T_o \tag{22}$$

with some $\varepsilon_2 > 0$. Combinning (17),(20) and (22) we can obtain

$$\varepsilon_2 \|\nabla u(t)\|_{m+2}^{m+2} \leq \|u\| \|u_t\| \leq C\|\nabla u\| \left(-\frac{d}{dt} J_o(u(t))\right)^{1/2}$$

or

$$\frac{d}{dt} J_o(u(t)) \leq -CJ_o(u(t))^{2(m+1)/(m+2)} \tag{23}$$

which implies

$$J_o(u(t)) \leq C(1+t)^{-(m+2)/m} \qquad \text{for } t \geq T_o \tag{24}.$$

Integrating (17) and using (24) we also have

$$\int_t^{\infty} \|u_t(s)\|_2^2 ds \leq C(1+t)^{-(m+2)/m}$$

Now, the desired estimate (8) has been established.

We proceed to the proof of (9). The case $N < m+2$ is trivial and we assume in the sequal that $N \geq m+2$. Let us return to the inequality (12). By Hölder's inequality

$$\int_{\Omega} |u|^{p+\alpha+2} dx \leq \|u\|_{p+2}^{\theta_1} \|u\|_{p_o+2}^{\theta_2} \|u\|_q^{\theta_3} \tag{25}$$

where $q=(p+m+2)N/(N-2-m)$ if $N > m+2$ and $q=\nu(p+m+2)$ if $N=m+2$ (ν being fixed sufficiently large number), and $\theta_i\,(i=1,2,3)$ are given by

$$\theta_1=(p+2)(1-\alpha q/r),\quad \theta_2=\alpha(1-mq/r) \text{ and } \theta_3=\alpha q(p+m+2)/r$$

with $r=m(q-p_o-2)+(p_o+2)(q-p-2)$.

Since $\|u(t)\|_{p_o+2}$ is bounded we have from (25) that

$$\int_{\Omega} |u|^{p+\alpha+2} dx \leq C_p \|u(t)\|_{p+2}^{p+2} + \frac{(p+1)(m+1)^{m+2}}{2(p+m+2)^{m+2}} \|\nabla u^{(p+m+2)/(m+2)}\|_{m+2}^{m+2}$$

where C_p is a constant satisfying

$$C_p \leq \text{const.}(1+p)^{\tilde{\theta}} \quad \text{with } \tilde{\theta} = \alpha q/(r-q).$$

Thus we arrive at the inequality

$$\frac{d}{dt}\|u(t)\|_{p+2}^{p+2} + C_2(1+p)^{-m} \|\nabla u^{(p+m+2)/(m+2)}\|_{m+2}^{m+2}$$

$$\leq C_3(1+p)^{\tilde{\theta}+1}\|u(t)\|_{p+2}^{p+2} \tag{26}$$

for any $p \geq 0$ and some $C_2, C_3 (>0)$ independent of p. Applying Lemma 3 to (26) we see that $\|u(t)\|_\infty \leq \text{const.} < \infty$ for any $t \geq 0$ under the additional assumption $u_0 \in L^\infty$.

Finally, setting $u(t)(1+t)^{1/m} \equiv w(t)$ and $\tau = \log(1+t)$, we have

$$\frac{d}{d\tau} w - \sum_{i=1}^{N} (|w_{x_i}|^m w_{x_i})_{x_i} = \frac{1}{m} w - \beta(x, e^{-\tau/m} w) . \tag{27}$$

Applying the argument obtaining the boundedness of $\|u(t)\|_\infty$ to (27) we can prove similarly the boundedness of $\|w(t)\|_\infty$, which shows (9).

3. Outline of the proof of Theorem 2.

It suffices again to derive the a priori estimates (10) and (11) for assumed smooth solutions.

Multiplying the equation (2) by u^{p+1} , $(u^{m+1})_t$ and u^{m+1} we have

$$\frac{1}{p+2}\frac{d}{dt}\|u(t)\|_{p+2}^{p+2} + \frac{4(m+1)(p+1)}{(p+m+2)^2}\|\nabla u^{(p+m+2)/2}\|_2^2$$

$$\leq k_0 \|u\|_{p+\alpha+2}^{p+\alpha+2} , \tag{28}$$

$$\frac{d}{dt} J_1(u(t)) + (m+1)\int_\Omega |u|^m |u_t|^2 \, dx = 0 \tag{29}$$

and

$$\|\nabla u^{m+1}(t)\|_2^2 \; - k_o \int_\Omega u^{m+\alpha+2} dx \;\leq \int_\Omega |u_t u^{m+1}| \, dx \qquad (30)$$

where we set

$$J_1(u) = \frac{1}{2}\|\nabla u^{m+1}\|_2^2 \; + (m+1)\int_\Omega \int_0^u \beta(x,s) s^m dsdx.$$

With the use of (28)-(30) we can prove the desired estimates in a parallel way to that of the previous section. The details of the proofs will be stated in [6].

References

1. N.D. Alikakos, L^p-bounds of solutions of reaction diffusion equations, Comm.Partial Differential Equations 4(8)1979) p.827-868.

2. V.A. Galaktinov, A boundary value problem for the nonlinear parabolic equation $u_t = \Delta u^{\alpha+1} + u^p$, Differential Equations 17(1981),p.551-555 (in Russian).

3. H.Ishii, Asymptotic stability and blowing up of solutions of some nonlinear evolution equations, J.Differential Equations,26(1977),p.291-319.

4. M.Nakao, On solutions of perturbed porous medium equations, Proc. 7th Conf.Ordin.& Part.Diff.Equat. at Dundee Univ., Lect.Notes of Math.,Springer,No.964(1982),p.539-547.

5. M.Nakao, L^p-estimates of solutions of some nonlinear degenerate diffusion equations, J.Math.Soc.Japan,to appear.

6. M.Nakao, Global solutions for some nonlinear parabolic equations with non-monotonic perturbations,in preparation.

7. M.Nakao & T.Narazaki, Existence and decay of solutions of some nonlinear parabolic variational inequalities, International J.Math.& Math.Sciences,Vol.2,No.1(1980), p.79-102.

8. M.Ôtani, Existence of strong solutions for $\frac{du}{dt} + \partial\psi^1(u) - \partial\psi^2(u) \ni f$, J.Fac.Sci.Univ.Tokyo,Sect. IA,Math.24(1977),p.575-605.

9. D.H. Sattinger, On global solutions of nonlinear hyperbolic equations, Arch.Rational Mech.Anal. 30(1968),p.148-172.

10. P.E. Sacks, Global behaviour for a class of nonlinear evolution equations, to appear.

11. M. Tsutsumi, Existence and nonexistence of global solutions for nonlinear parabolic equations,Pub.R.I.M.S.,Kyoto Univ. 8 (1972-73),p.211-229.

Lecture Notes in Num. Appl. Anal., 8, 175–228 (1985)
Recent Topics in Nonlinear PDE II, Sendai, 1985

On a Global Existence Theorem of Neumann Problem for Some Quasi-Linear Hyperbolic Equations

Yoshihiro SHIBATA* and Yoshio TSUTSUMI**

* Department of Mathematics, University of Tsukuba
Ibaraki 305, Japan

** Department of Pure and Applied Sciences, College of General Education,
University of Tokyo, Tokyo 113, Japan

1. Problem, Notations and Results

Let Ω be an unbounded domain in an n dimensional Euclidean space $\mathbb{R}^n$ with compact and C^∞ boundary $\partial\Omega$. Let us denote a time variable by t or x_0 and space variables by $x = (x_1,\dots,x_n)$, respectively. For differentiation we write $\partial_t = \partial_0 = \partial/\partial t$, $\partial_j = \partial/\partial x_j$, $j = 1,\dots,n$. In this note, we consider the following problem:

$$
\begin{aligned}
&\Phi(\Lambda u) = \Box u - \Sigma_{ij=1}^n \partial_i(\alpha_{ij}(D_x^1 u)\partial_j u) + \beta(D^1 u) = f && \text{in } [0,\infty)\times\Omega,\\
(1.1)\quad &\Phi_D(\tilde{\Lambda} u) = \frac{\partial u}{\partial\nu} + \Sigma_{ij=1}^n \nu_i(x)\alpha_{ij}(D_x^1 u)\partial_j u + \gamma(u) = g && \text{on } [0,\infty)\times\partial\Omega,\\
&u(0,x) = u_0(x),\ (\partial_t u)(0,x) = u_1(x) && \text{in } \Omega.
\end{aligned}
$$

Here, $D_x^1 u = (\partial_i u;\ i = 1,\dots,n)$, $D^1 u = (\partial_0 u,\ D_x^1 u)$, $\tilde{\Lambda} u = (D_x^1 u, u)$, $\Lambda u = (D^1 u,\ \partial_i\partial_j u,\ i,j = 0,1,\dots,n)$, $\Box = \partial_t^2 - \Delta = \partial_t^2 - \Sigma_{j=1}^n \partial_j^2$ (Δ: Laplacian, $\Box$: d'Alembertian), $\frac{\partial}{\partial\nu} = \Sigma_{j=1}^n \nu_j(x)\partial_j$, and $\nu_i(x)$, $i = 1,\dots,n$, are real-valued functions in $C^\infty(\mathbb{R}^n)$ such that $\nu(x) = (\nu_1(x),\dots,\nu_n(x))$ represents the unit outer normal of $\partial\Omega$ at $x \in \partial\Omega$.

The typical examples of Φ and Φ_b are the following:

$$\Phi(\Lambda u) = \partial_t^2 u - \Sigma_{j=1}^n \partial_j(\partial_j u(1 + \Sigma_{k=1}^n (\partial_k u)^2)^{-\frac{1}{2}}),$$

(1.2)

$$\Phi_b(\lambda u) = \Sigma_{j=1}^n \nu_j(x)\partial_j u(1 + \Sigma_{k=1}^n (\partial_k u)^2)^{-\frac{1}{2}}.$$

In fact, since by Taylor expansion we can write

$$\partial_j u(1+\Sigma_{k=1}^n(\partial_k u)^2)^{-\frac{1}{2}} = \partial_j u - \tfrac{1}{2}\partial_j u(\Sigma_{k=1}^n(\partial_k u)^2)\int_0^1 [1+a(\Sigma_{k=1}^n(\partial_k u)^2)]^{-\frac{3}{2}} da$$

we have

$$(1.3)\quad \alpha_{ij}(D_x^1) = -\tfrac{1}{2}\,\delta_{ij}\int_0^1 [1 + a(\Sigma_{k=1}^n(\partial_k u)^2)]^{-\frac{3}{2}}\, da\, (\Sigma_{k=1}^n (\partial_k u)^2),$$

where $\delta_{ij} = 1$ if $i = j$ and $= 0$ if $i \neq j$.

It is the purpose of this note that we prove that there exists one and only one classical solution of (1.1) for sufficiently small and smooth data u_0, u_1, f and g. First, we summarize notations.

Notations. n and N, N', N", ℓ always refer to the dimension and the integers defined in Theorem 1.3 below. K, L, M and p, q always refer to non-negative integers and extended real numbers with $1 \leq p, q \leq \infty$, respectively. Functions considered below are all real-valued. For an integer $s \geq 1$, a function u, a vector valued function $f = (f_1,\dots,f_s)$ and a multi-index $\alpha = (\alpha_1,\dots,\alpha_n)$, set

$$\partial_x^\alpha = \partial_1^{\alpha_1}\cdots\partial_n^{\alpha_n},\ |\alpha| = \alpha_1+\cdots+\alpha_n,\ |f| = |f_1| +\cdots+ |f_s|,$$

$$\overline{D}_x^L D_x^M u = (\partial_x^\alpha u;\ M \leq |\alpha| \leq M+L),\ \overline{D}^L D^M u = (\partial_t^j\partial_x^\alpha u;\ M \leq j+|\alpha| \leq M+L),$$

$$\overline{D}_x^L D_x^M f = (\overline{D}_x^L D_x^M f_1,\dots,\overline{D}_x^L D_x^M f_s),\ \overline{D}^L D^M f = (\overline{D}^L D^M f_1,\dots,\overline{D}^L D^M f_s),$$

$$D_x^0 u = D^0 u = u \quad (\overline{D}_x^L D_x^M u,\ \overline{D}_x^L D_x^M f,\dots \text{ are all vector valued functions}).$$

In particular, $\Lambda u = \bar{D}^1 D^1 u$ and $\chi u = \bar{D}_x^1 u$. Let G and I be domains in $\mathbb{R}^n$ and $\mathbb{R}^s$, respectively. Let X be a Banach space. $L^p(G)$ denotes the usual L^p space over G and $\|\cdot\|_{G,p}$ its usual norm. Put $\|f\|_{G,p} = \Sigma_{j=1}^s \|f_j\|_{G,p}$. Set

$$H_p^L(G) = \{u \in L^p(G);\ \|u\|_{G,p,L} = \|\bar{D}_x^L u\|_{G,p} < \infty\}.$$

$C^L(I;X)$ denotes the space of functions which are L-times continuously differentiable for $t \in I$ with range in X. $C^L(\bar{I};X)$ and $C_0^L(I;X)$ denote the spaces of functions in $C^L(I;X)$ all of whose derivatives of order $\leq L$ have continuous extensions to $\bar{I}$ and with compact support in I, respectively. Put

$$C^L(I) = C^L(I;\mathbb{R}^1),\ C_0^L(I) = C_0^L(I;\mathbb{R}^1),\ C^L(\bar{I}) = C^L(\bar{I};\mathbb{R}^1),$$

$$\mathcal{B}^L(\bar{I}) = \{u \in C^L(\bar{I});\ \|u\|_{I,\infty,L} < \infty\},\ E_p^L(G) = \cap_{j=0}^L C^j([0,\infty);\ H_p^{L-j}(G)).$$

$\mathcal{S} = \mathcal{S}(\mathbb{R}^n)$ denotes the Schwartz class. Let A be a set and $g = (g_1,\dots,g_s)$ a vector. If $g_j \in A$, $j = 1,\dots,s$, we write $g \in A$. For $f = (f_1,\dots,f_s) \in E_p^L(G)$ and $u \in E_p^L(G)$ and a real number k, put

$$|u|_{G,p,L,k} = \sup_{t \geq 0} (1+t)^k \|\bar{D}^L u(t,\cdot)\|_{G,p},\ |f|_{G,p,L,k} = \Sigma_{j=1}^s |f_j|_{G,p,L,k},$$

$$\mathbb{E}_p^L(G) = \{u \in E_p^L(G);\ (\partial_t^j u)(0,x) = 0,\ j = 0,1,\dots,L\},$$

$$E_p^{L,k}(G) = \{u \in E_p^L(G);\ |u|_{G,p,L,k} < \infty\},\ \mathbb{E}_p^{L,k}(G) = \mathbb{E}_p^L(G) \cap E_p^{L,k}(G).$$

R always refers to a real number such that $\mathbb{R}^n - \Omega \subset \{x \in \mathbb{R}^n;\ |x| \leq R-1\}$. For any $r \geq R-1$, put $\Omega_r = \{x \in \Omega;\ |x| < r\}$. For simplicity, we write

$$\|u\|_p = \|u\|_{\Omega,p},\ \|u\| = \|u\|_2,\ \|u\|_{p,L} = \|u\|_{\Omega,p,L},\ \|u\|_p^R = \|u\|_{\Omega_R,p},$$

$$\|u\|^R = \|u\|_2^R,\ |u|_{p,L,k} = |u|_{\Omega,p,L,k},\ |u|_{p,L,k}^R = |u|_{\Omega_R,p,L,k},$$

$\|u\|^R_{p,L} = \|u\|_{\Omega_R,p,L}$ (for $f = (f_1,\dots,f_s)$, we use the same abbreviation),

$E^L_p = E^L_p(\Omega)$, $E^{L,k}_p = E^{L,k}_p(\Omega)$, $\mathbb{E}^L_p = \mathbb{E}^L_p(\Omega)$, $\mathbb{E}^{L,k}_p = \mathbb{E}^{L,k}_p(\Omega)$,

$E^L_{p,R} = E^L_p(\Omega_R)$, $E^{L,k}_{p,R} = E^{L,k}_p(\Omega_R)$, $\mathbb{E}^L_{p,R} = \mathbb{E}^L_p(\Omega_R)$, $\mathbb{E}^{L,k}_{p,R} = \mathbb{E}^{L,k}_p(\Omega_R)$.

For any real number k, [k] is the "integral-part" function used in number theory. δ_{ij} denotes Kronecker's delta (cf. (1.3)). Put

$$(u,v) = \int_\Omega u(x)v(x)dx, \ \langle u,v\rangle = \int_{\partial\Omega} u(y)v(y)dS_y, \ \langle u\rangle^2 = \langle u,u\rangle,$$

where dS_y denotes the surface element of $\partial\Omega$. Let us define the special constants depending on dimension n as follows:

$$p(n) = \begin{cases} 4, & n \geq 6, \\ 6, & 3 \leq n \leq 5, \end{cases} \qquad d(n) = \begin{cases} (n-1)/4, & n \geq 6, \\ 4/3, & n = 5, \\ 1, & n = 4, \\ 4/7, & n = 3. \end{cases}$$

$$q(n) = \begin{cases} 4/3, & n \geq 6, \\ 6/5, & 3 \leq n \leq 5, \end{cases}$$

$$r(n) = \begin{cases} 1, & n \geq 6, \\ 2, & 3 \leq n \leq 5, \end{cases}$$

For a non-linear function $H(t,x,y_1,\dots,y_s)$, put

$$(d^LH)(t,x,y_1,\dots,y_s)(z^1_1,\dots,z^1_s)\cdots(z^L_1,\dots,z^L_s)$$

$$= [\frac{\partial^L}{\partial a_1\cdots\partial a_L} H(t,x,(y_1,\dots,y_s) + \Sigma^L_{j=1} a_j(z^j_1,\dots,z^j_s))]\Big|_{a_1=\cdots=a_L=0}.$$

Let λ_j, $j = 0,\dots,n$, λ_{n+1} and λ_{ij}, $i,j = 0,\dots,n$ be independent variables corresponding to $\partial_j u$, $j = 0,\dots,n$, u and $\partial_i\partial_j u$, $i,j = 0,\dots,n$, respectively. Put $\lambda = (\lambda_i,\ i = 0,\dots,n,\ \lambda_{ij},\ i,j = 0,\dots,n)$, $\lambda' = (\lambda_0,\dots,\lambda_n)$, $\lambda'' = (\lambda_1,\dots,\lambda_n)$, $\tilde{\lambda} = (\lambda'',\lambda_{n+1})$. In the course of calculations below various constants will be simply denoted by C. Especially C(*⋯*) denotes a constant depending on quantities appearing in parentheses.

Special constants will be denoted by c_0, c_1,... etc.

Now, we introduce our assumptions.

Assumption 1.1. 1° The spacial dimension $n \geq 3$.

2° α_{ij}, β and γ are all real-valued functions such that

$$\alpha_{ij} \in \mathcal{B}^\infty(\{\lambda'' \in \mathbb{R}^n; |\lambda''| \leq 1\}),\ \beta \in \mathcal{B}^\infty(\{\lambda' \in \mathbb{R}^{n+1}; |\lambda'| \leq 1\}),$$

$$\gamma \in \mathcal{B}^\infty(\{\lambda_{n+1} \in \mathbb{R}^1; |\lambda_{n+1}| \leq 1\}),$$

$$\alpha_{ij}(\lambda'') = \begin{cases} O(|\lambda''|) & \text{near } \lambda'' = 0 \quad \text{if } n \geq 6, \\ O(|\lambda''|^2) & \text{near } \lambda'' = 0 \quad \text{if } 3 \leq n \leq 5, \end{cases}$$

$$\beta(\lambda') = \begin{cases} O(|\lambda'|^2) & \text{near } \lambda' = 0 \quad \text{if } n \geq 6, \\ O(|\lambda'|^3) & \text{near } \lambda' = 0 \quad \text{if } 3 \leq n \leq 5, \end{cases}$$

$$\gamma(\lambda_{n+1}) = \begin{cases} O(|\lambda_{n+1}|^2) & \text{near } \lambda_{n+1} = 0 \quad \text{if } n \geq 6, \\ O(|\lambda_{n+1}|^3) & \text{near } \lambda_{n+1} = 0 \quad \text{if } 3 \leq n \leq 5, \end{cases}$$

3° Ω is non-trapping in the following sense: Let $G(t,x,y)$ be the Green function of the following problem:

$$[(\frac{\partial}{\partial t})^2 - \Sigma_{j=1}^n(\frac{\partial}{\partial x_j})^2]G(t,x,y) = 0 \text{ in } (0,\infty)\times\Omega,\quad \frac{\partial G}{\partial \nu}\Big|_{x\in\partial\Omega} = 0,$$

$$\lim_{t\to 0+} \partial_t^j G = \begin{cases} 0, & j = 0, \\ \delta(x-y), & j = 1, \end{cases} \quad \text{in } \Omega,$$

where y is an arbitrary point in Ω and δ is the Dirac delta function. Let a and b be positive numbers such that $R-1 \leq a \leq b$. For any $v \in L^2(\Omega)$ with $\text{supp } v \subset \Omega_a$, put

$$(Gv)(t,x) = \int_\Omega G(t,x,y)v(y)dy.$$

Then, there exists a positive constant T depending only on a, b, n and Ω such that $(Gv)(t,x) \in C^\infty([T,\infty)\times\overline{\Omega_b})$.

Remark 1.2. 1° The above definition of "non-trapping" is due to Vainberg [15].

2° If the complement of Ω is convex, then Ω is non-trapping (see e.g. Melrose [4]).

3° It follows from (1.3) that the operator defined in (1.2) satisfies the assumption 2°.

The purpose of this note is to prove the following.

Theorem 1.3. *Assume that Assumption* 1.1 *is satisfied. Let* ℓ *be a non-negative integer and put*

$$N'' = \max(\ell+1,\ 2n+[n/p(n)]+8),\quad N' = 2N''+1,\quad N = N'+2n+[n/p(n)]+8.$$

Then, there exist positive numbers δ_0 *and* δ_1 *depending on* N, α_{ij}, β, γ, n *and* Ω *such that if data* u_0, u_1, f *and* g *satisfy the following conditions:*

$$(1.4)\quad \begin{aligned} &u_0 \in \mathcal{B}^{N+1}(\bar{\Omega}) \cap H^{N+1}_{q(n)}(\Omega),\ u_1 \in \mathcal{B}^{N}(\bar{\Omega}) \cap H^{N}_{q(n)}(\Omega),\ g \in E_2^{N-2,d(n)r(n)},\\ &f \in \mathcal{B}^{N-1}([0,\infty)\times\bar{\Omega}) \cap E^{N-1}_{q(n)} \cap E^{N-4,d(n)r(n)}_{q(n)} \cap E_2^{N'+1,d(n)r(n)}, \end{aligned}$$

u_0, u_1, f *and* g *satisfy the* $N-1$ *th order compatibility condition* (*cf. section* V *below*),

$$(1.5)\quad \begin{aligned} &\|u_0\|_{\infty,N+1} + \|u_0\|_{q(n),N+1} + \|u_1\|_{\infty,N} + \|u_1\|_{q(n),N} + \|\bar{D}^{N-1}f(0,\cdot)\|_{\infty} +\\ &\|\bar{D}^{N-1}f(0,\cdot)\|_{q(n)} + |f|_{2,N'+1,d(n)r(n)} + |f|_{q(n),N-4,d(n)r(n)} +\\ &|g|_{2,N-2,d(n)r(n)} \leq \delta_1\cdot\delta \end{aligned}$$

for some δ *with* $0 < \delta \leq \delta_0$, *then there exists a solution* $u \in C^{\ell+2}([0,\infty)\times\bar{\Omega})$ *of* (1.1) *with*

$$|\Lambda u|_{2,\ell,0} + |\Lambda u|_{p(n),\ell,d(n)} + |\Lambda u|_{\infty,\ell,d(n)} + |u|^R_{2,0,0} +$$
$$|u|^R_{p(n),0,d(n)} + |u|^R_{\infty,0,d(n)} \leqq \delta.$$

Furthermore, there exists a small number $\delta_2 > 0$ such that if u and v are solutions in $C^2([0,\infty)\times\bar{\Omega})$ of (1.1) for the same data such that

$$|\Lambda u|_{\infty,0,0} + |u|^R_{\infty,0,0} \leqq 1, \; |\Lambda v|_{\infty,0,0} + |v|^R_{\infty,0,0} \leqq 1, \; |D^1_x u|_{\infty,0,0} \leqq \delta_2,$$

then u = v.

Remark 1.4. If u is a solution of (1.1), it is necessary that u satisfies the conditions: $|(\Lambda u)(t,x)| \leq 1$ for all $(t,x) \in [0,\infty)\times\bar{\Omega}$ and $|u(t,x)| \leqq 1$ for all $(t,x) \in [0,\infty)\times\partial\Omega$ (cf. Assumption 1.1- 2°).

Our proof of the existence theorem for (1.1) is a straightforward adaptation of a quadratic iteration scheme with a process of "smoothing" which is well-known as the Nash-Moser technique. We explain the reason why we use the Nash-Moser technique. For this, let us consider the case where $\alpha_{ij} = \beta = 0$ in (1.1). In the usual sense, the linearized problem is the following:

$$\Box u = f \text{ in } [0,\infty)\times\Omega, \quad \frac{\partial u}{\partial \nu} = g \text{ on } [0,\infty)\times\partial\Omega,$$
(1.6)
$$u(0,x) = u_0(x), \; (\partial_t u)(0,x) = u_1(x) \text{ in } \Omega.$$

For simplicity, we assume that $u_0 = u_1 \equiv 0$. According to a result due to Miyatake [5] we have

(1.7) $$\|D^1 u(t,\cdot)\| \leqq Ce^{\sigma t}\int_0^t \{\|f(s,\cdot)\|^2 + \|D^1_x g(s,\cdot)\|^2\}\, ds,$$

where u is a solution of (1.6) and C and σ are some positive constants depending on n and Ω. To our knowledge, we can not prove the global existence theorem for (1.1) by using (1.7) because of the term: $e^{\sigma t}$.

On the other hand, by the well-known energy method (cf. section III below) we have

$$(1.8)\quad \|D^1u(t,\cdot)\| \leqq C \int_0^t \{ \|f(s,\cdot)\| + \|\bar{D}^2 g(s,\cdot)\|^R \}^2 \, ds.$$

This estimate is good for the constant (the term $e^{\sigma t}$ disappears). But, we encounter the derivative loss because of the term: $\|\bar{D}^2 g(t,\cdot)\|^R$. To overcome the difficulties caused by the derivative loss of this type, the Nash-Moser technique is very useful and the proof is easy and short as compared with other methods. From this point of view, we adopt the Nash-Moser technique to prove the global existence theorem for (1.1).

2. Preliminaries

2.1.- SMOOTHING OPERATOR. Choose $\psi(t) \in S(\mathbb{R}^1)$ and $\phi(x) \in S(\mathbb{R}^n)$ so that

$$(2.1.1)\quad 0 \leqq \psi \leqq 1,\ \operatorname{supp}\psi \subset [0,\infty), \int_{-\infty}^{\infty} \psi(t)dt = 1,\ \int_{-\infty}^{\infty} \psi(t)t^j dt = 0,\ j \geqq 1,$$

$$(2.1.2)\quad 0 \leqq \phi \leqq 1,\ \int_{\mathbb{R}^n} \phi(x)dx = 1,\ \int_{\mathbb{R}^n} \phi(x)x^\alpha\, dx = 0,\ |\alpha| \geqq 1.$$

In fact, the existence of functions satisfying the properties (2.1.1) follows immediately from Boas' theorem (cf. Widder [17]). If ϕ is the inverse Fourier transformation of a function in $C_0^\infty(\mathbb{R}^n)$ which is 1 near the origin, ϕ satisfies (2.1.2). For any $u \in \mathbb{E}_p^L$, put $u_0 = u$ if $t \geqq 0$ and $= 0$ if $t < 0$. Since the boundary of Ω is C^∞ and compact, via local map, by using well-known Seeley's extension method, we have that there exists a function $u'(t,x)$ defined on $\mathbb{R}^1 \times \mathbb{R}^n$ satisfying the following properties:

$$u' \in \cap_{j=0}^{L} C^j(\mathbb{R}^1;\, H_p^{L-j}(\mathbb{R}^n)), \quad u' = u_0 \ \text{ on } \mathbb{R}^1\times\Omega,$$

(2.1.3)

$$\|\bar{D}^K u'(t,\cdot)\|_{\mathbb{R}^n,p} \leq C(K,p)\, \|\bar{D}^K u(t,\cdot)\|_p, \qquad t > 0,\ K \leq L.$$

For any real number $\tau \geq 1$, let us define our smoothing operator $S(\tau)u$ by

(2.1.4) $$S(\tau)u = \int_{\mathbb{R}^{n+1}} \tau^{n+1}\psi((t-s)\tau)\phi((x-y)\tau)u'(s,y)dsdy.$$

In the same way as in Shibata [9, Lemma 11.1], by using (2.1.1), (2.1.2) and (2.1.3) we have

Lemma 2.1.1. *Let k be a non-negative real number. Then, for any $u \in \mathbf{E}_p^{L,K} \cap C^L([0,\infty)\times\bar{\Omega})$, the following assertions are valid.*

(a) $S(\tau)u \in \mathbf{E}_p^{\infty,k}$,

(b) $|S(\tau)u|_{p,M,k} \leq C(p,M,k)|u|_{p,L,k}$, $\quad 0 \leq M \leq L$,

(c) $|S(\tau)u|_{p,M,k} \leq C(p,M,L,k)\tau^{M-L}|u|_{p,L,k}$, $\quad 0 \leq L \leq M$,

(d) $|(I-S(\tau))u|_{p,M,k} \leq C(p,M,L,k)\tau^{M-L}|u|_{p,L,k}$, $\quad 0 \leq M \leq L$,

where I denotes the identity operator.

2.2- INTERPOLATION INEQUALITIES. In the paragraphs 2.2 and 2.3, D and k, m always refer to a domain with C^∞ and compact boundary and non-negative real numbers, respectively.

Lemma 2.2.1 (see Shibata [8, Lemma 2.2.7]). *The following interpolation inequalities hold.*

(i) $\|u\|_{D,p,M} \leq C(D,p,L)\, \|u\|_{D,p}^{1-(M/L)}\, \|u\|_{D,p,L}^{M/L}$, $\quad u \in H_p^L(D),\ 0 \leq M \leq L,$

(ii) $|v|_{D,p,M,k} \leq C(D,p,L,k)|v|_{D,p,L,k}^{1-(M/L)}|v|_{D,p,L,k}^{M/L}$, $\quad v \in E_p^{L,k}(D),\ 0 \leq M \leq L.$

By Lemma 2.2.1 and Young's inequality, we have

Lemma 2.2.2 (see Shibata [8, Lemma 2.2.9]). *The following three inequalities hold.*

(a) $$|u|_{D,p,K,k}|v|_{D,q,M,m} \leq C(p,q,K,M,k,m,D)[|u|_{D,p,K+M,k}|v|_{D,q,0,m} + |u|_{D,p,0,k}|v|_{D,q,K+M,m}],\ u \in E_p^{K+M,k}(D),\ v \in E_q^{K+M,m}(D).$$

(b) $$\|f\|_{D,p,K}\|g\|_{D,q,M} \leq C(p,q,K,M,D)[\|f\|_{D,p,K+M}\|g\|_{D,q} + \|f\|_{D,p}\|g\|_{D,q,K+M}],\ f \in H_p^{K+M}(D),\ g \in H_q^{K+M}(D).$$

(c) $$|u|_{D,p,K,k}\|g\|_{D,q,M} \leq C(p,q,K,M,k,D)[|u|_{D,p,K+M,k}\|g\|_{D,q} + |u|_{D,p,0,k}\|g\|_{D,q,K+M}],\ u \in E_p^{K+M,k}(D),\ g \in H_q^{K+M}(D).$$

2.3- *ESTIMATES OF COMPOSED FUNCTIONS AND PRODUCTS OF FUNCTIONS.*

Lemma 2.3.1 (see Klainerman [3, Lemma 5.1]). *Let* $G(x,y_1,\dots,y_s)$ *be a function in* $\mathcal{B}^\infty(\overline{D}\times\{y = (y_1,\dots,y_s) \in \mathbb{R}^s\ ;\ |y| \leq 1\})$ *such that* $G(x,0) = 0$. *Let* $u(x) = (u_1(x),\dots,u_s(x)) \in H_p^L(D)$ *such that* $\|u\|_\infty \leq 1$. *Then,*

$$\|G(\cdot,u(\cdot))\|_{D,p,L} \leq C(p,L,G,D)\ \|u\|_{D,p,L}.$$

Lemma 2.3.2 (see Shibata [8, Lemma 5.12]). *Let* $H(t,x,y_1,\dots,y_s)$ *be a function in* $\mathcal{B}^\infty([0,\infty)\times\overline{D}\times\{y = (y_1,\dots,y_s) \in \mathbb{R}^s\ ;\ |y| \leq 1\})$. *Let* $u(t,x) = (u_1(t,x),\dots,u_s(t,x)) \in \mathcal{B}^L([0,\infty)\times\overline{D})$ *such that* $|u|_{D,\infty,0,0} \leq 1$. *Then,*

$$|H(\cdot,\cdot,u(\cdot,\cdot))|_{D,\infty,L,0} \leq C(L,H)(1 + |u|_{D,\infty,L,0}).$$

Lemma 2.3.3. *Let* H *be the same as in Lemma* 2.3.2. *In addition, assume that* $H(t,x,0) = 0$. *Let* $u(t,x) = (u_1(t,x),\dots,u_s(t,x)) \in E_p^{L,k}(D) \cap$

$E^L_\infty(D)$ such that $|u|_{D,\infty,0,0} \leq 1$. Then,

$$|H(\cdot,\cdot,u(\cdot,\cdot))|_{D,p,L,k} \leq C(p,L,k)[|u|_{D,p,L,k} + |u|_{D,p,0,k}|u|_{D,\infty,L,0}].$$

Proof. By Taylor expansion, we have

$$(2.3.1)\quad H(t,x,u(t,x)) = \Sigma_{j=1}^{s} \; G_j(t,x,u(t,x))u_j(t,x),$$

where $G_j(t,x,u(t,x)) = \int_0^1 \frac{\partial H}{\partial y_j}(t,x,au(t,x))da$. Applying Leibniz's formula to (2.3.1), we have

$$(2.3.2)\quad |D^K H(\cdot,\cdot,u(\cdot,\cdot))|_{D,p,0,k} \leq$$

$$C(L)\Sigma_{j=1}^{s} \; \Sigma_{M=0}^{K} \; |G_j(\cdot,\cdot,u(\cdot,\cdot))|_{D,\infty,M,0}|u_j|_{D,p,K-M,k}, \; 0 \leq K \leq L.$$

Applying Lemmas 2.2.2 and 2.3.2 to (2.3.2), we obtain the lemma.

Lemma 2.3.4. Let s be an integer ≥ 2, k, $k_1,\ldots,k_s$ non-negative real numbers such that $k = k_1+\cdots+k_s$, p, $p_1,\ldots,p_s$ extended real numbers such that $1 \leq p$, $p_1,\ldots,p_s \leq \infty$ and $p^{-1} = \Sigma_{j=1}^{s} \; p_j^{-1}$. Then, for any $u_j \in E_{p_j}^{L,k_j}(D)$ the inequality:

$$|u_1\cdots u_s|_{D,p,L,k} \leq C\Sigma_{i=1}^{s}[|u_i|_{D,p_i,L,k_i} \prod_{j\neq i} |u_j|_{D,p_j,0,k_j}]$$

holds with some constant $C = C(p,L,k,s,D) > 0$.

Proof. By Leibniz's formula and Hölder's inequality, we have

$$(2.3.3)\quad \begin{aligned} &|u_1\cdots u_s|_{D,p,L,k} \leq \\ &C(L) \sum_{j_1+\cdots+j_s=L} |u_1|_{D,p_1,j_1,k_1}\cdots|u_s|_{D,p_s,j_s,k_s}. \end{aligned}$$

Applying Lemma 2.2.2 to (2.3.3), we obtain the lemma.

2.4- SOME INEQUALITIES FOR THE TRACE OPERATOR.

Lemma 2.4.1. (see e.g. Mizohata [6, Chapter 3]). *If* $u \in H_2^1(\Omega)$, *then for any* $r \geq R-1$ *the inequality*: $< u > \leq C(n,\Omega,r) \|u\|_{\Omega_r,2,1}$ *holds*.

Lemma 2.4.2. *If* $n \geq 3$ *and* $u \in H_2^1(\Omega)$, *then for any* $r \geq R-1$ *we have the inequality*: $\|u\|_{\Omega_r,2} \leq C(n,\Omega,r) \|D_x^1 u\|$.

Lemma 2.4.2 can be proved by using Rellich's compactness theorem (see e.g. Mizohata [6, Theorem 3.3]) and the identity:

$$r^{-n+2} \Sigma_{i=1}^n [\partial_i (r^{(n-2)/2} u)]^2 = [(\tfrac{n-2}{2})^2 + \tfrac{n-2}{2}] u^2 r^{-2} +$$

$$\Sigma_{j=1}^n (\partial_j u)^2 + \frac{n-2}{2} \frac{\partial}{\partial r}(r^{-1} u^2) \quad (r = |x|,\ n \geq 3).$$

For further details, Heywood [1] can be referred. Combining Lemmas 2.4.1 and 2.4.2, we have

Lemma 2.4.3. *Assume that* $n \geq 3$. *Then, there exists a constant* c_0 *depending only on* n *and* Ω *such that for any* $u \in H_2^1(\Omega)$ *the inequality*: $< u > \leq c_0 \|D_x^1 u\|$ *holds*.

2.5- SOME ELLIPTIC ESTIMATES. In this paragraph, we give some " a priori" estimates of solutions of the following equation:

$$(2.5.1)\quad \Delta u + \Sigma_{ij=1}^n \partial_i (a_{ij} \partial_j u) = f \text{ in } \Omega, \quad \frac{\partial u}{\partial \nu} + \Sigma_{ij=1}^n \nu_i a_{ij} \partial_j u + bu = g \text{ on } \partial\Omega.$$

Lemma 2.5.1. *Assume that*

$$a_{ij} \in \mathcal{B}^{L+1}(\overline{\Omega}), \quad b \in \mathcal{B}^{L+1}(\overline{\Omega}), \ \Sigma_{ij=1}^n \|a_{ij}\|_{\infty,1} + \|b\|_\infty \leq 1,$$

$$f \in H_2^L(\Omega), \quad g \in H_2^{L+1}(\Omega),$$

Then, there exists a constant $c_{1,L} > 0$ *depending only on* Ω, n *and* L *such that if* $\Sigma_{ij=1}^{n} \|a_{ij}\|_\infty \leq c_{1,L}$, *then for any solution* $u \in H_2^{L+2}(\Omega)$ *of* (2.5.1) *the following estimates hold.*

$$\|D_x^2 u\|_{2,K} \leq C(K,n,\Omega)[\ \|f\|_{2,K} + \|g\|_{2,K+1}^R + \|D_x^1 u\|_{2,K} +$$

$$(\Sigma_{ij=1}^{n} \|a_{ij}\|_{\infty,K+1} + \|b\|_{\infty,K+1}) \|D_x^1 u\|\ ],\ 0 \leq K \leq L.$$

In order to prove Theorem 2.5.1, we need the following two well-known results

Lemma 2.5.2. *Let* $f \in H_2^L(\Omega)$, $g \in H_2^{L+1}(\Omega)$ *and* u *be a solution of the equation:* $\Delta u = f$ *in* Ω, $\frac{\partial u}{\partial \nu} = g$ *on* $\partial\Omega$. *Then, we have*

$$\|\overline{D}_x^{K+2} u\|_{\Omega_{R-(1/2)},2} \leq C(n,\Omega,R,L)[\ \|f\|_{2,K} + \|g\|_{2,K+1}^R + \|u\|_{2,1}^R],\ 0 \leq K \leq L.$$

Lemma 2.5.3. *There exists a constant* $C = C(L) > 0$ *such that for any* $v \in H_2^{L+2}(\mathbb{R}^n)$

$$\|D_x^2 v\|_{\mathbb{R}^n,2,K} \leq C\ \|\Delta v\|_{\mathbb{R}^n,2,K}, \qquad 0 \leq K \leq L.$$

Recall that $\mathbb{R}^n - \Omega \subset \{x \in \mathbb{R}^n;\ |x| < R-1\}$. Choose $\phi(x) \in C^\infty(\mathbb{R}^n)$ so that $\phi(x) = 1$ if $|x| \geq R-(1/2)$ and $= 0$ if $|x| \leq R-1$. If $u \in H_2^{L+2}(\Omega)$ satisfies the equation (2.5.1), we have

$$(2.5.2) \qquad \Delta(\phi u) = \phi F + 2\Sigma_{j=1}^{n} \partial_j\phi \cdot \partial_j u + \Delta\phi \cdot u \quad \text{in } \mathbb{R}^n$$

where $F = f - \Sigma_{ij=1}^{n} \partial_i(a_{ij}\partial_j u)$. Since supp $\partial_j\phi$, supp $\Delta\phi \subset \{x \in \mathbb{R}^n;\ R-1 \leq |x| \leq R-(1/2)\}$, applying Lemma 2.5.2 to (2.5.2), we have

$$(2.5.3) \qquad \|D_x^2 u\|_{E_{R-(1/2)},2,K} \leq C(L,R)[\ \|F\|_{2,K} + \|D_x^1 u\|_{2,K} + \|u\|^R]$$

where $E_{R-(1/2)} = \{x \in \mathbb{R}^n;\ |x| \geq R-(1/2)\}$. On the other hand, if we put

$G = -\Sigma_{ij=1}^n \nu_i a_{ij}\partial_j u - bu + g$, u satisfies

$$(2.5.4)\quad \Delta u = F \text{ in } \Omega,\ \frac{\partial u}{\partial\nu} = G \text{ on } \partial\Omega.$$

Applying Lemma 2.5.2 to (2.5.4), we have

$$(2.5.5)\quad \|D_x^2 u\|_{\Omega_{R-(1/2)},2,K} \leqq C(L,n,\Omega)[\ \|F\|_{2,K} + \|G\|_{2,K+1}^R + \|D_x^1 u\| + \|u\|^R].$$

Combining (2.5.3) and (2.5.5) and using Lemma 2.4.2, we have

$$(2.5.6)\quad \|D_x^2 u\|_{2,K} \leqq C(L,n,\Omega)[\ \|F\|_{2,K} + \|G\|_{2,K+1}^R + \|D_x^1 u\|_{2,K}].$$

By Lemmas 2.2.2 and 2.4.2 and Leibniz's formula, we have

$$(2.5.7)\quad \begin{aligned}\|F\|_{2,K} \leq{}& \|f\|_{2,K} + (\Sigma_{ij=1}^n \|a_{ij}\|_\infty)\ \|D_x^{K+2}u\| + \\ & C(L)[\ \|D_x^1 u\|_{2,K} + (\Sigma_{ij=1}^n \|a_{ij}\|_{\infty,K+1})\ \|D_x^1 u\|\ ],\end{aligned}$$

$$(2.5.8)\quad \begin{aligned}\|G\|_{2,K+1}^R \leqq{}& \|g\|_{2,K+1}^R + (\Sigma_{ij=1}^n \|a_{ij}\|_\infty)(\Sigma_{j=1}^n \|\nu_j\|_\infty)\ \|D_x^{K+2}u\| + \\ & C(R,L,\Omega)[\ \|D_x^1 u\|_{2,K} + (\Sigma_{ij=1}^n \|a_{ij}\|_{\infty,K+1} + \|b\|_{\infty,K+1})\ \|D_x^1 u\|\ ].\end{aligned}$$

Choose a positive constant $c_{1,L}$ so small that $c_{1,L}(1 + \Sigma_{j=1}^n \|\nu_j\|_\infty)C(L,n,\Omega) \leqq 1/2$, $C(L,n,\Omega)$ being the constant in (2.5.6). Substituting (2.5.7) and (2.5.8) into (2.5.6), we have the theorem.

3. L^2-Estimates for Linearized Equations.

In this section, we give L^2-estimates (energy estimates) of solutions of the following hyperbolic equation:

$$(3.1)\quad \begin{aligned}&\partial_t^2 u - \Sigma_{ij=1}^n \partial_i((\delta_{ij} + a_{ij})\partial_j u) + \Sigma_{j=0}^n b_j^1\partial_j u = f && \text{in } [0,\infty)\times\Omega,\\ &\Sigma_{ij=1}^n \nu_i(\delta_{ij} + a_{ij})\partial_j u + b^0 u = g && \text{on } [0,\infty)\times\partial\Omega,\end{aligned}$$

$$u(0,x) = (\partial_t u)(0,x) = 0 \qquad \text{in } \Omega.$$

Theorem 3.1. *Put* $A = (a_{ij}, b_j^1, b^0)$. *Let* L *be an integer* ≥ 4. *Assume that*

1° $A \in B^L([0,\infty)\times\bar{\Omega}) \cap E_\infty^{L,d(n)r(n)}$,

2° $|A|_{\infty,2,d(n)r(n)} \leq 1$, $|b^0|_{\infty,0,0} \leq (8c_0^2)^{-1}$, $\Sigma_{ij=1}^n |a_{ij}|_{\infty,0,0} \leq c_{1,L}$,

where c_0 *and* $c_{1,L}$ *are the same as in Lemmas* 2.4.3 *and* 2.5.1, *respectively*,

3° $\delta_{ij} + a_{ij}(t,x) = \delta_{ji} + a_{ji}(t,x)$, $\Sigma_{ij=1}^n(\delta_{ij} + a_{ij}(t,x))\xi_i\xi_j \geq \frac{1}{2}|\xi|^2$

for all $\xi = (\xi_1,\ldots,\xi_n) \in \mathbb{R}^n$ *and* $(t,x) \in [0,\infty)\times\bar{\Omega}$,

4° $f \in E_2^L \cap E_2^{L-1,d(n)r(n)}$, $g \in E_2^{L+1,d(n)r(n)}$,

$$(\partial_t^j f)(0,x) = (\partial_t^j g)(0,x)\Big|_{x\in\partial\Omega} = 0, \quad 0 \leq j \leq L-1.$$

Then, there exists a unique solution $u \in \mathbb{E}_2^{L+1}$ *of* (3.1) *which has the following estimates*:

$$|D^1 u|_{2,M,0} \leq C(M,n,\Omega)[|f|_{2,M,d(n)r(n)} + |g|^R_{2,M+2,d(n)r(n)} + |A|_{\infty,M+1,d(n)r(n)}\{|f|_{2,0,d(n)r(n)} + |g|^R_{2,2,d(n)r(n)}\}],$$

for any integer M *with* $0 \leq M \leq L-1$.

Remark 3.2. If $A \in B^\infty([0,\infty)\times\bar{\Omega})$, according to Miyatake [5], we have that there exists a unique solution of (3.1). But, if A belongs to at least $B^4([0,\infty)\times\bar{\Omega})$, estimates of Miyatake's type for (3.1) can be obtained. So, by the limiting process we obtain the unique existence theorem for (3.1) under the assumptions 1°-4° (see Shibata [10]).

Proof of Theorem 3.1. Noting that $u \in \mathbb{E}_2^{L+1}$ and that $0 \leqq M \leqq L-1$ and differentiating (3.1) M times with respect to t, we have

$$\partial_t^2(\partial_t^M u) - \Delta(\partial_t^M u) - \Sigma_{ij=1}^n \partial_i[\Sigma_{k=0}^M \binom{M}{k} \partial_t^k a_{ij} \partial_t^{M-k} \partial_j u] + \Sigma_{j=0}^n \Sigma_{k=0}^M \binom{M}{k} \partial_t^k b_j^1 \partial_t^{M-k} \partial_j u = \partial_t^M f \quad \text{in } [0,\infty)\times\Omega,$$

$$(3.3)\quad \frac{\partial}{\partial\nu}(\partial_t^M u) + \Sigma_{ij=1}^n \nu_i[\Sigma_{k=0}^M \binom{M}{k} \partial_t^k a_{ij} \partial_t^{M-k} \partial_j u] = \partial_t^M g - \Sigma_{k=0}^M \binom{M}{k} \partial_t^k b^0 \partial_t^{M-k} u \quad \text{on } [0,\infty)\times\partial\Omega,$$

$$(\partial_t^M u)(0,x) = (\partial_t^{M+1} u)(0,x) = 0 \quad \text{in } \Omega.$$

By integration by parts, we have the identity:

$$-(\Delta(\partial_t^M u) + \Sigma_{ij=1}^n \partial_i[\Sigma_{k=0}^M \binom{M}{k} \partial_t^k a_{ij} \partial_t^{M-k} \partial_j u], \partial_t^{M+1} u) =$$

$$\frac{d}{dt}[<\partial_t^M g, \partial_t^M u> - \frac{1}{2}<b^0 \partial_t^M u, \partial_t^M u> - \Sigma_{k=1}^M \binom{M}{k} <\partial_t^k b^0 \partial_t^{M-k} u, \partial_t^M u> +$$

$$\frac{1}{2}\Sigma_{ij=1}^n((\delta_{ij} + a_{ij})\partial_t^M \partial_i u, \partial_t^M \partial_j u) + \Sigma_{ij=1}^n \Sigma_{k=1}^M \binom{M}{k} (\partial_t^k a_{ij} \partial_i \partial_t^{M-k} u, \partial_j \partial_t^M u)]$$

$$(3.4)\quad - <\partial_t^{M+1} g, \partial_t^M u> + \frac{3}{2}<\partial_t b^0 \partial_t^M u, \partial_t^M u> + \Sigma_{k=2}^M \binom{M}{k} <\partial_t^k b^0 \partial_t^{M+1-k} u, \partial_t^M u> +$$

$$\Sigma_{k=1}^M \binom{M}{k} <\partial_t^{k+1} b^0 \partial_t^{M-k} u, \partial_t^M u> - \frac{3}{2}\Sigma_{ij=1}^n (\partial_t a_{ij} \partial_t^M \partial_i u, \partial_t^M \partial_j u) -$$

$$\Sigma_{ij=1}^n \Sigma_{k=2}^M \binom{M}{k} (\partial_t^k a_{ij} \partial_t^{M+1-k} \partial_i u, \partial_t^M \partial_j u) -$$

$$\Sigma_{ij=1}^n \Sigma_{k=1}^M \binom{M}{k} (\partial_t^{k+1} a_{ij} \partial_t^{M-k} \partial_i u, \partial_t^M \partial_j u).$$

Multiplying (3.3) by $\partial_t^{M+1} u$ and integrating over $[0,t]\times\Omega$, by (3.4) and the assumption 3° we have

$$(3.5)\quad 4^{-1} \| D^1 \partial_t^M u(t,\cdot)\|^2 \leqq$$

$$|<\partial_t^M g(t,\cdot), \partial_t^M u(t,\cdot)>| + \frac{1}{2}\| b^0(t,\cdot)\|_\infty <\partial_t^M u(t,\cdot)>^2 +$$

$$\Sigma_{k=1}^{M}\binom{M}{k}\|\partial_t^k b^0(t,\cdot)\|_\infty\ |<\partial_t^{M-k}u(t,\cdot),\ \partial_t^M u(t,\cdot)>|\ +$$

$$\Sigma_{ij=1}^{n}\Sigma_{k=1}^{M}\binom{M}{k}\|\partial_t^k a_{ij}(t,\cdot)\|_\infty\cdot|(\partial_i\partial_t^{M-k}u(t,\cdot),\partial_j\partial_t^M u(t,\cdot))|\ +$$

$$\int_0^t|<\partial_t^{M+1}g(s,\cdot),\ \partial_t^M u(s,\cdot)>|ds + \frac{3}{2}\int_0^t\|\partial_t b^0(s,\cdot)\|_\infty <\partial_t^M u(s,\cdot)>^2\ ds\ +$$

$$\Sigma_{k=1}^{M}\binom{M}{k}\int_0^t\|\partial_t^{k+1}b^0(s,\cdot)\|_\infty <\partial_t^{M-k}u(s,\cdot)><\partial_t^M u(s,\cdot)>\ ds\ +$$

$$\Sigma_{k=2}^{M}\binom{M}{k}\int_0^t\|\partial_t^k b^0(s,\cdot)\|_\infty <\partial_t^{M+1-k}u(s,\cdot)><\partial_t^M u(s,\cdot)>\ ds\ +$$

$$\frac{3}{2}\Sigma_{ij=1}^{n}\int_0^t\|\partial_t a_{ij}(s,\cdot)\|_\infty\|\partial_t^M\partial_i u(s,\cdot)\|\cdot\|\partial_t^M\partial_j u(s,\cdot)\|\ ds\ +$$

$$\Sigma_{ij=1}^{n}\Sigma_{k=2}^{M}\binom{M}{k}\int_0^t\|\partial_t^k a_{ij}(s,\cdot)\|_\infty\|\partial_t^{M+1-k}\partial_i u(s,\cdot)\|\cdot\|\partial_t^M\partial_j u(s,\cdot)\|\ ds\ +$$

$$\Sigma_{ij=1}^{n}\Sigma_{k=1}^{M}\binom{M}{k}\int_0^t\|\partial_t^{k+1}a_{ij}(s,\cdot)\|_\infty\|\partial_t^{M-k}\partial_i u(s,\cdot)\|\cdot\|\partial_t^M\partial_j u(s,\cdot)\|\ ds\ +$$

$$\Sigma_{j=0}^{n}\Sigma_{k=0}^{M}\binom{M}{k}\int_0^t\|\partial_t^k b_j^1(s,\cdot)\|_\infty\|\partial_t^{M-k}\partial_j u(s,\cdot)\|\cdot\|\partial_t^{M+1}u(s,\cdot)\|\ ds\ +$$

$$\int_0^t\|\partial_t^M f(s,\cdot)\|\ \|\partial_t^{M+1}u(s,\cdot)\|\ ds.$$

We are going to evaluate each term. For simplicity, let us put

$$I = (|D^1u|_{2,M-1,0} + |A|_{\infty,M+1,d(n)r(n)}|D^1u|_{2,0,0})^2,$$

$$J(t) = \int_0^t (1+s)^{-d(n)r(n)}\|D^1\partial_t^M u(s,\cdot)\|^2\ ds.$$

By the assumption 2°, Lemmas 2.4.1, 2.4.3 and 2.2.2, we have

(3.6.a) $$|<\partial_t^M g(t,\cdot),\ \partial_t^M u(t,\cdot)>| + \frac{1}{2}\|b^0(t,\cdot)\|_\infty <\partial_t^M u(t,\cdot)>^2 \leqq$$

$$((16)^{-1}+(48)^{-1})\|D_x^1\partial_t^M u(t,\cdot)\|^2 + C(M)(\|\bar{D}_x^{1}\partial_t^M g(t,\cdot)\|^R)^2 \leqq$$

$$((16)^{-1}+(48)^{-1})\|D_x^1\partial_t^M u(t,\cdot)\|^2 + C(M)(|g|^R_{2,M+1,d(n)r(n)})^2,$$

(3.6.b) $$\Sigma_{k=1}^{M}\binom{M}{k}\|\partial_t^k b^0(t,\cdot)\|_\infty\ |<\partial_t^{M-k}u(t,\cdot),\ \partial_t^M u(t,\cdot)>| \leqq$$

$$\Sigma_{k=1}^{M}\binom{M}{k} c_0^2 \|\partial_t^k b^0(t,\cdot)\|_\infty \|D_x^1\partial_t^{M-k}u(t,\cdot)\|\cdot\|D_x^1\partial_t^M u(t,\cdot)\| \leq$$

$$C(M)\ \Sigma_{k=1}^{M} |\partial_t b^0|_{\infty,k-1,0}|D^1u|_{2,M-k,0}\|D_x^1\partial_t^M u(s,\cdot)\| \leq$$

$$C(M)(|D^1u|_{2,M-1,0} + |A|_{\infty,M,0}|D^1u|_{2,0,0})\|D_x^1\partial_t^M u(t,\cdot)\| \leq$$

$$C(M)I + (48)^{-1}\|D_x^1\partial_t^M u(t,\cdot)\|^2 \quad \text{(this term disappears when } M = 0\text{)},$$

(3.6.c) $$\Sigma_{ij=1}^{n}\Sigma_{k=1}^{M}\binom{M}{k}\|\partial_t^k a_{ij}(s,\cdot)\|_\infty\cdot|(\partial_i\partial_t^{M-k}u(t,\cdot),\ \partial_j\partial_t^M u(t,\cdot))| \leq$$

$$C(M)I + (48)^{-1}\|D_x^1\partial_t^M u(t,\cdot)\|^2 \quad \text{(this term disappears when } M = 0\text{)},$$

(3.6.d) $$\int_0^t |\langle\partial_t^{M+1}g(s,\ ),\ \partial_t^M u(s,\cdot)\rangle|\,ds \leq$$

$$C(M)\int_0^t (1+s)^{d(n)r(n)}(\|\partial_t^{M+1}g(s,\cdot)\|_{2,1}^R)^2\,ds + \int_0^t (1+s)^{-d(n)r(n)}\|D_x^1\partial_t^M u(s,\cdot)\|^2\,ds \leq$$

$$C(M)(|g|_{2,M+2,d(n)r(n)}^R)^2\int_0^t (1+s)^{-d(n)r(n)}\,ds + J(t) \leq$$

$$C(M)(|g|_{2,M+2,d(n)r(n)}^R)^2 + J(t),$$

(3.6.e) $$\frac{3}{2}\int_0^t \|\partial_t b^0(s,\cdot)\|_\infty \langle\partial_t^M u(s,\cdot)\rangle^2\,ds \leq$$

$$\frac{3}{2}c_0^2\,|A|_{\infty,1,d(n)r(n)}\int_0^t (1+s)^{-d(n)r(n)}\|D_x^1\partial_t^M u(s,\cdot)\|^2\,ds \leq$$

$$\frac{3}{2}c_0^2\,J(t),$$

(3.6.f) $$\Sigma_{k=1}^{M}\binom{M}{k}\int_0^t \|\partial_t^{k+1}b^0(s,\cdot)\|_\infty \langle\partial_t^{M-k}u(s,\cdot)\rangle\langle\partial_t^M u(s,\cdot)\rangle\,ds \leq$$

$$C(M)\Sigma_{k=1}^{M}\int_0^t (1+s)^{d(n)r(n)}\|\partial_t^{k+1}b^0(s,\cdot)\|^2\|D_x^1\partial_t^{M-k}u(s,\cdot)\|^2\,ds + \int_0^t (1+s)^{-d(n)r(n)}\|D_x^1\partial_t^M u(s,\cdot)\|^2\,ds \leq$$

$$C(M)\Sigma_{k=1}^{M}|\partial_t^2 b^0|_{\infty,k-1,d(n)r(n)}^2|D^1u|_{2,M-k,0}^2\int_0^t (1+s)^{-d(n)r(n)}ds + J(t)$$

$$\leq C(M)I + J(t) \qquad \text{(this term disappears when } M = 0\text{)},$$

(3.6.g) $$\Sigma_{k=2}^{M}\binom{M}{k}\int_0^t \|\partial_t^k b^0(s,\cdot)\|_\infty <\partial_t^{M+1-k}u(s,\cdot)><\partial_t^M u(s,\cdot)> ds \leq$$

$$C(M)(\Sigma_{k=2}^{M} |\partial_t^2 b^0|_{\infty,k-2,d(n)r(n)}|D^1 u|_{2,M+1-k,0})^2 +$$

$$\int_0^t (1+s)^{-d(n)r(n)} \|D_x^1\partial_t^M u(s,\cdot)\|^2 ds \leq$$

$$C(M)I + J(t) \qquad \text{(this term disappears when } 0 \leq M \leq 1\text{)},$$

(3.6.h) $$\frac{3}{2}\Sigma_{ij=1}^{n}\int_0^t \|\partial_t a_{ij}(s,\cdot)\|_\infty \|\partial_t^M\partial_i u(s,\cdot)\|\cdot\|\partial_t^M\partial_j u(s,\cdot)\| ds \leq$$

$$\frac{3}{2}|A|_{\infty,1,d(n)r(n)}\int_0^t (1+s)^{-d(n)r(n)} \|D_x^1\partial_t^M u(s,\cdot)\|^2 ds \leq \frac{3}{2}J(t),$$

(3.6.i) $$\Sigma_{ij=1}^{n}\Sigma_{k=2}^{M}\binom{M}{k}\int_0^t \|\partial_t^k a_{ij}(s,\cdot)\|_\infty \|\partial_t^{M+1-k}\partial_i u(s,\cdot)\|\cdot\|\partial_t^M\partial_j u(s,\cdot)\| ds \leq$$

$$C(M)\Sigma_{ij=1}^{n}(\Sigma_{k=2}^{M} |\partial_t^2 a_{ij}|_{\infty,k-2,d(n)r(n)}|D_x^1 u|_{2,M+1-k,0})^2 +$$

$$\int_0^t (1+s)^{-d(n)r(n)} \|D_x^1\partial_t^M u(s,\cdot)\|^2 ds \leq$$

$$C(M)I + J(t) \qquad \text{(this term disappears when } 0 \leq M \leq 1\text{)},$$

(3.6.j) $$\Sigma_{ij=1}^{n}\Sigma_{k=1}^{M}\binom{M}{k}\int_0^t \|\partial_t^{k+1} a_{ij}(s,\cdot)\|_\infty \|\partial_t^{M-k}\partial_i u(s,\cdot)\|\cdot\|\partial_t^M\partial_j u(s,\cdot)\| ds \leq$$

$$C(M)I + J(t) \qquad \text{(this term disappears when } 0 \leq M \leq 1\text{)},$$

(3.6.k) $$\Sigma_{j=0}^{n}\Sigma_{k=0}^{M}\binom{M}{k}\int_0^t \|\partial_t^k b_j^1(s,\cdot)\|_\infty \|\partial_t^{M-k}\partial_j u(s,\cdot)\|\cdot\|\partial_t^{M+1}u(s,\cdot)\| ds +$$

$$\int_0^t \|\partial_t^M f(s,\cdot)\|\cdot\|\partial_t^{M+1}u(s,\cdot)\| ds \leq$$

$$C(M)I + 2J(t) + C|f|_{2,M,d(n)r(n)}^2,$$

where for notational convenience the same letter $C(M)$ is used to denote constants depending essentially on M, n and Ω.

Combining (3.5) and (3.6), by Gronwall's inequality we have

$$(3.7)\quad \begin{aligned}\|D^1\partial_t^M u(t,\cdot)\| &\le C(n,M,\Omega)[|f|_{2,M,d(n)r(n)} + |g|^R_{2,M+2,d(n)r(n)} + \\ &\eta(M)\{|D^1u|_{2,M-1,0} + |A|_{\infty,M+1,d(n)r(n)}|D^1u|_{2,0,0}\}]\end{aligned}$$

for all M with $0 \leqq M \leqq L-1$, where $\eta(M)$ is a constant such that $\eta(M) = 0$ if $M = 0$ and $= 1$ if $M \geqq 1$. In particular, we have (3.2) when $M = 0$.

Now, we shall prove (3.2) by induction on M. We may assume that $1 \leqq M \leqq L-1$ and that (3.2) are already proved for smaller values of M. We shall prove

$$(3.8)\quad \begin{aligned}|D_x^K\partial_t^{M+1-K}u|_{2,0,0} &\leqq C(n,M,\Omega)[|f|_{2,M,d(n)r(n)} + |g|^R_{2,M+2,d(n)r(n)} + \\ &|A|_{\infty,M+1,d(n)r(n)}\{|f|_{2,0,d(n)r(n)} + |g|^R_{2,2,d(n)r(n)}\}].\end{aligned}$$

We prove (3.8) also by induction on K. For simplicity, let us put

$$B = |D^1u|_{2,M-1,0} + |A|_{\infty,M+1,0}|D^1u|_{2,0,0}.$$

Applying the induction hypotheses to (3.7), we have that (3.8) are valid for K = 0 and 1. We may assume that $2 \leqq K \leqq M+1$ and that (3.8) are already proved for smaller values of K. Differentiating (3.1) M+1-K times with respect to t, we have

$$(3.9)\quad \begin{aligned}-\Delta(\partial_t^{M+1-K}u) - \Sigma_{ij=1}^n \partial_i(a_{ij}\partial_j\partial_t^{M+1-K}u) &= F_K \qquad \text{in } \Omega,\\ \frac{\partial}{\partial\nu}(\partial_t^{M+1-K}u) + \Sigma_{ij=1}^n \nu_i a_{ij}\partial_j\partial_t^{M+1-K}u &= G_K \qquad \text{on } \partial\Omega,\end{aligned}$$

where

$$(3.10)\quad \begin{aligned}F_K = \partial_t^{M+1-K}f - \partial_t^{M+3-K}u + \Sigma_{ij=1}^n\Sigma_{k=1}^{M+1-K}\binom{M+1-K}{k}\{\partial_t^k a_{ij}\partial_i\partial_j\partial_t^{M+1-K-k}u + \\ \partial_i\partial_t^k a_{ij}\partial_j\partial_t^{M+1-K-k}u\} - \Sigma_{j=0}^n\Sigma_{k=0}^{M+1-K}\binom{M+1-K}{k}\partial_t^k b_j^1\partial_t^{M+1-K-k}\partial_j u,\end{aligned}$$

$$G_K = \partial_t^{M+1-K} g - \Sigma_{ij=1}^n \nu_i \Sigma_{k=1}^{M+1-K} \binom{M+1-K}{k} \partial_t^k a_{ij} \partial_t^{M+1-K-k} \partial_j u -$$

$$\Sigma_{k=0}^{M+1-K} \binom{M+1-K}{k} \partial_t^k b^0 \partial_t^{M+1-K-k} u.$$

By the assumption 2°, we have $\Sigma_{ij=1}^n |a_{ij}|_{\infty,0,0} \leq c_{1,L}$, which implies that we can apply Theorem 2.5.1 to (3.9). Thus, we have

$$(3.11)\quad \begin{aligned} &|D_x^K \partial_t^{M+1-K} u|_{2,0,0} \leq C(n,L,\Omega)[|F_K|_{2,K-2,0} + |G_K|_{2,K-1,0}^R + \\ &\quad |D_x^1 \partial_t^{M+1-K} u|_{2,K-2,0} + |A|_{\infty,K-1,0} |D_x^1 \partial_t^{M+1-K} u|_{2,0,0}]. \end{aligned}$$

By Lemmas 2.3.6, 2.2.2 and 2.4.2, Leibniz's formula and the assumption 2°, we have

$$(3.12.a)\quad \begin{aligned} &\Sigma_{ij=1}^n \Sigma_{k=1}^{M+1-K} \binom{M+1-K}{k} |\partial_t^k a_{ij} \partial_i \partial_j \partial_t^{M+1-K-k} u|_{2,K-2,0} \leq \\ &\Sigma_{ij=1}^n \Sigma_{k=1}^{M+1-K} \binom{M+1-K}{k} [|\partial_t^k a_{ij}|_{\infty,K-2} |\partial_i \partial_j \partial_t^{M+1-K-k} u|_{2,0,0} + \\ &\qquad |\partial_t^k a_{ij}|_{\infty,0,0} |\partial_i \partial_j \partial_t^{M+1-K-k} u|_{2,K-2,0}] \leq \\ &C(M) \Sigma_{k=1}^{M+1-K} [|\partial_t a_{ij}|_{\infty,K+k-3} |D_x^1 u|_{2,M+2-K-k,0} + \\ &\qquad |\partial_t a_{ij}|_{\infty,k-1,0} |D_x^1 u|_{2,M-k,0}] \leq C(M)B, \end{aligned}$$

$$(3.12.b)\quad \Sigma_{ij=1}^n \Sigma_{k=1}^{M+1-K} \binom{M+1-K}{k} |\partial_i \partial_t^k a_{ij} \partial_j \partial_t^{M+1-K-k} u|_{2,K-2,0} \leq C(M)B,$$

$$(3.12.c)\quad \Sigma_{j=0}^n \Sigma_{k=0}^{M+1-K} \binom{M+1-K}{k} |\partial_t^k b_j^1 \partial_t^{M+1-K-k} \partial_j u|_{2,K-2,0} \leq C(M)B,$$

$$(3.12.d)\quad \begin{aligned} &\Sigma_{k=1}^{M+1-K} \binom{M+1-K}{k} |\partial_t^k a_{ij} \partial_t^{M+1-K-k} \partial_j u|_{2,K-1,0}^R \leq \\ &C(M) \Sigma_{k=1}^{M+1-K} [|\partial_t^k a_{ij}|_{\infty,K-1,0}^R |\partial_t^{M+1-K-k} \partial_j u|_{2,0,0}^R + \\ &\qquad |\partial_t^k a_{ij}|_{\infty,0,0}^R |\partial_t^{M+1-K-k} \partial_j u|_{2,K-1,0}^R] \leq \\ &C(M) \Sigma_{k=1}^{M+1-K} [|\partial_t a_{ij}|_{\infty,K+k-2,0}^R |\partial_j u|_{2,M+1-K-k,0}^R + \end{aligned}$$

$$|\partial_t a_{ij}|^R_{\infty,k-1,0}|\partial_j u|^R_{2,M-k,0}] \leq C(M)B,$$

$$(3.12.e)\quad \Sigma_{k=0}^{M+1-K}\binom{M+1-K}{k}|\partial_t^k b^0 \partial_t^{M+1-K-k} u|^R_{2,K-1,0} \leq C(M)B,$$

$$(3.12.f)\quad |A|_{\infty,K-1,0}|D_x^1 \partial_t^{M+1-K} u|_{2,0,0} \leq \qquad (K \geq 2)$$

$$C|\bar{D}^1 A|_{\infty,K-2,0}|D_x^1 u|_{2,M+1-K,0} \leq C(M)B,$$

where for notational convenience the same letter $C(M)$ is used to denote constants depending essentially on M, n and Ω.

Combining (3.10), (3.11) and (3.12), we have

$$|D_x^K \partial_t^{M+1-K} u|_{2,0,0} \leq$$

$$(3.13)\quad C(M,n,\Omega)[|f|_{2,M-1,0} + |D_x^{K-2}\partial_t^{M+3-K} u|_{2,0,0} + |g|^R_{2,M,0} +$$

$$|D^1 u|_{2,M-1,0} + |A|_{\infty,M+1,0}|D^1 u|_{2,0,0}].$$

Applying the induction hypotheses to (3.13), we have (3.8), which completes the proof.

4. Uniform Decay Estimates.

In this section, we give the uniform decay estimates of solutions of the equation (3.1).

Theorem 4.1. Let L be an integer $\geq 2n+5$ and u a solution of (3.1). In addition to all assumptions of Theorem 3.1, assume that

1° $n \geq 3$ and Ω is non-trapping,

2° $A \in E^{L-2,d(n)r(n)}_{p(n)/r(n)}$, $|A|_{p(n)/r(n),1,d(n)r(n)} \leq 1$,

3° $f \in E^{L-3,d(n)r(n)}_{q(n)}$.

Then, for any integer M with $0 \le M \le L-2n-5$ we have

$$|\bar{D}^1 u|^R_{p(n),0,d(n)} \le C[|f|_{q(n),2n+2,d(n)r(n)} +$$

$$(4.1) \qquad |g|^R_{2,2n+4,d(n)r(n)} + |D^1 u|_{2,2n+4,0} +$$

$$\{|A|_{p(n)/r(n),2n+3,d(n)r(n)} + |A|_{\infty,2n+4,d(n)r(n)}\}|D^1 u|_{2,0,0}],$$

$$|D^1 u|_{p(n),M,d(n)} \le C(M)[|f|_{q(n),M+2n+2,d(n)r(n)} +$$

$$(4.2) \qquad |g|^R_{2,M+2n+4,d(n)r(n)} + |D^1 u|_{2,M+2n+4,0} +$$

$$\{|A|_{p(n)/r(n),M+2n+3,d(n)r(n)} + |A|_{\infty,M+2n+4,d(n)r(n)}\}|D^1 u|_{2,0,0}].$$

Before proving the theorem, we discuss the decay rate of solutions of the Neumann problem for the wave equation:

$$\Box u = f \text{ in } [0,\infty)\times\Omega, \quad \frac{\partial u}{\partial \nu} = g \text{ on } [0,\infty)\times\partial\Omega,$$

(4.3)

$$u(0,x) = u_0(x), \quad (\partial_t u)(0,x) = u_1(x) \text{ in } \Omega.$$

For simplicity, assume that u is a C^∞ solution of (4.3) and that all norms for u_0, u_1, f and g appearing below are finite. If $g = 0$, then we have

$$|\bar{D}^1 u|^R_{p(n),0,d(n)} \le C(n,\Omega,R)[\ \|u_0\|_{q(n),2n+3} +$$

(4.4)

$$\|u_1\|_{q(n),2n+2} + |f|_{q(n),2n+2,d(n)r(n)}],$$

$$|D^1 u|_{p(n),L,d(n)} \le C(n,\Omega,L)[\ \|u_0\|_{q(n),L+2n+3} +$$

(4.5)

$$\|u_1\|_{q(n),L+2n+2} + |f|_{q(n),L+2n+2,d(n)r(n)}].$$

The proofs of (4.4) and (4.5) are given in Appendix. When $g \neq 0$, we can represent u as follows: Let v_1 be a solution of the equation:

$$-\Delta v_1 + \mu v_1 = 0 \text{ in } \Omega_R, \ \frac{\partial v_1}{\partial \nu} = g(t,\cdot) \text{ on } \partial\Omega, \ \frac{\partial v_1}{\partial r} = 0 \text{ on } \{x \in \mathbb{R}^n ; |x| = R\},$$

where $r = |x|$, t is regarded as a parameter and μ is a sufficiently large positive number. Choose $\phi(x) \in C_0^\infty(\mathbb{R}^n)$ so that $\phi(x) = 1$ if $|x| \leqq R-(1/2)$ and $= 0$ if $|x| \geqq R$ and put

$$v(t,x) = \phi(x)v_1(t,x) \text{ if } x \in \Omega_R \text{ and } = 0 \text{ if } x \notin \Omega_R.$$

Let w be a solution of the Neumann problem:

$$\Box w = f - \Box v \text{ in } [0,\infty)\times\Omega, \ \frac{\partial w}{\partial \nu} = 0 \text{ on } [0,\infty)\times\partial\Omega,$$

$$w(0,x) = u_0(x) - v(0,x), \ (\partial_t w)(0,x) = u_1(x) - (\partial_t w)(0,x) \text{ in } \Omega.$$

Then, by the uniqueness theorem for (4.3) we have $u \doteqdot v + w$. Since by the usual elliptic estimate in an interior domain we have

$$|v_1|^R_{2,K,d(n)r(n)} \leqq C(K,n,\Omega)|g|^R_{2,K,d(n)r(n)}$$

and since $1 \leqq q(n) \leqq 2$, by Sobolev's inequality, (4.4) and (4.5) we have

$$(4.6) \quad |\bar{D}^1 u|^R_{p(n),0,d(n)} \leqq C[\, \|u_0\|_{q(n),2n+3} + \|u_1\|_{q(n),2n+2} + |f|_{q(n),2n+2,d(n)r(n)} + |g|^R_{2,2n+4,d(n)r(n)}],$$

$$(4.7) \quad |D^1 u|_{p(n),L,d(n)} \leqq C[\, \|u_0\|_{q(n),L+2n+3} + \|u_1\|_{q(n),L+2n+2} + |f|_{q(n),L+2n+2,d(n)r(n)} + |g|^R_{2,L+2n+4,d(n)r(n)}].$$

<u>Proof of Theorem 4.1.</u> Let u be a solution of (3.1). Then, we can regard u as a solution of the equation:

$$(4.8) \quad \Box u = f + \Sigma^n_{ij=1} \partial_i(a_{ij}\partial_j u) - \Sigma^n_{j=0} b^1_j \partial_j u \qquad \text{in } [0,\infty)\times\Omega,$$

$$\frac{\partial u}{\partial \nu} = g-\Sigma_{ij=1}^{n}\nu_i a_{ij}\partial_j u - b^0 u \text{ on } [0,\infty)\times\partial\Omega,\ u\big|_{t=0} = \partial_t u\big|_{t=0} = 0 \text{ in } \Omega.$$

Since $q(n)^{-1} = 2^{-1} + (p(n)/r(n))^{-1}$, it follows from Lemmas 2.3.6, 2.2.2 and 2.4.2 with some constant $C = C(K,n,\Omega)$ that

$$|b^0u|^R_{2,K+2n+4,d(n)r(n)} \leq$$

(4.9.a) $$C[|b^0|^R_{\infty,K+2n+4,d(n)r(n)}|u|^R_{2,0,0} + |u|^R_{2,K+2n+4,0}] \leq$$

$$C[|D^1u|_{2,K+2n+3,0} + |A|_{\infty,K+2n+4,d(n)r(n)}|D^1u|_{2,0,0}],$$

$$|\partial_i(a_{ij}\partial_j u)|_{q(n),K+2n+2,d(n)r(n)} \leq$$

$$C[|a_{ij}|_{p(n)/r(n),K+2n+2,d(n)r(n)}|\partial_i\partial_j u|_{2,0,0} +$$

(4.9.b) $$|a_{ij}|_{p(n)/r(n),0,d(n)r(n)}|\partial_i\partial_j u|_{2,K+2n+2,0} +$$

$$|\partial_i a_{ij}|_{p(n)/r(n),K+2n+2,d(n)r(n)}|\partial_j u|_{2,0,0} +$$

$$|\partial_i a_{ij}|_{p(n)/r(n),0,d(n)r(n)}|\partial_j u|_{2,K+2n+2,0}] \leq$$

$$C[|D^1u|_{2,K+2n+3,0} + |A|_{p(n)/r(n),K+2n+3,d(n)r(n)}|D^1u|_{2,0,0}],$$

(4.9.c) $$|b^1_j\partial_j u|_{q(n),K+2n+2,d(n)r(n)} \leq$$

$$C[|D^1u|_{2,K+2n+2,0} + |A|_{p(n)/r(n),K+2n+2,d(n)r(n)}|D^1u|_{2,0,0}],$$

(4.9.d) $$|a_{ij}\partial_j u|^R_{2,K+2n+4,d(n)r(n)} \leq$$

$$C[|D^1u|_{2,K+2n+4,0} + |A|_{\infty,K+2n+4,d(n)r(n)}|D^1u|_{2,0,0}].$$

Thus, applying (4.6) and (4.7) to (4.8) and using (4.9), we have the theorem.

5. Compatibility Condition and Reduction of Problem.

In this section, we shall introduce the definition of compatibility condition and reduce the problem (1.1) to the case where $u_0 = u_1 = 0$. For the first purpose, we introduce some notations:

$$\alpha'_{ij}(\lambda_1,\ldots,\lambda_n) = \alpha_{ij}(\lambda_1,\ldots,\lambda_n) + \Sigma_{k=1}^{n}(\partial_{\lambda_j}\alpha_{ik})(\lambda_1,\ldots,\lambda_n)\lambda_k,$$

$$(5.1)\quad \tilde{\Phi}(\Lambda u) = -\Sigma_{ij=1}^{n}\partial_i(\alpha_{ij}(D_x^1 u)\partial_j u) + \beta(D^1 u),$$

$$\tilde{\Phi}_b(D_x^1 u) = \Sigma_{ij=1}^{n}\nu_i\alpha_{ij}(D_x^1 u)\partial_j u.$$

It is obvious that

$$(5.2)\quad \partial_k(\Sigma_{j=1}^{n}\alpha_{ij}(D_x^1 u)\partial_j u) = \Sigma_{j=1}^{n}\alpha'_{ij}(D_x^1 u)\partial_j\partial_k u,\quad k = 0,1,\ldots,n.$$

Let us define functions G^k, $k \geqq 0$, by the formula:

$$(5.3)\quad \partial_t^k(\tilde{\Phi}(\Lambda u)) = G^k(\bar{D}_x^2 u,\ldots,\bar{D}_x^2\partial_t^k u,\partial_t^{k+1}u).$$

Let us define successively u_{k+2}, $k \geqq 0$, by

$$(5,4)\quad u_{k+2} = \Delta u_k + G^k(\bar{D}_x^2 u_0,\ldots,\bar{D}_x^2 u_k,u_{k+1}) + (\partial_t^k f)(0,x).$$

By Assumption 1.1-2° we have that there exist functions π_k, $k\geq 0$, such that

$$(5.5)\quad \pi_k(0) = 0,\ u_{k+2} = \pi_k(\bar{D}_x^{k+2}u_0(x),\ \bar{D}_x^{k+1}u_1(x),\ (\bar{D}^k f)(0,x)).$$

Next, let us define vector valued functions $\Phi_b^k(x,\cdot)$ by

$$(5.6)\quad \begin{aligned}\partial_t^k(\Phi_b(\tilde{\Lambda}u)) = {} & \Sigma_{ij=1}^{n}\nu_i(x)(\delta_{ij} + \alpha'_{ij}(D_x^1 u))\partial_j\partial_t^k u + \\ & \gamma'(u)\partial_t^k u + \Phi_b^k(x,\tilde{\Lambda}u,\ldots,\tilde{\Lambda}\partial_t^{k-1}u).\end{aligned}$$

By Assumption 1.1- 2°, we have $\Phi_b^k(x,0) = 0$ for all $k \geqq 0$.

Now, we introduce the definition of the compatibility condition.

Definition 5.1. *Let L be an integer ≥ 2. We say that data u_0, u_1, f and g for (1.1) satisfy the L-th order compatibility condition if*

$$\Sigma_{ij=1}^{n} \nu_i(\delta_{ij} + \alpha_{ij}(D_x^1 u_0))\partial_j u_0 + \gamma(u_0) = g(0,x), \ x \in \partial\Omega,$$

$$\Sigma_{ij=1}^{n} \nu_i(\delta_{ij} + \alpha_{ij}'(D_x^1 u_0))\partial_j u_k + \gamma'(u_0)u_k +$$

$$\Phi_b^k(x,\tilde{\lambda}u_0,\ldots,\tilde{\lambda}u_{k-1}) = (\partial_t^k g)(0,x), \ x \in \partial\Omega, \ k = 1,2,\ldots,L.$$

Now, we reduce the problem (1.1) to the case where $u_0 = u_1 = 0$. In the same way as in Hörmander [2, Lemma 5.6.1], we have

Lemma 5.2. *Let $p_1,\ldots,p_m$ be extended real numbers with $1 \leq p_i \leq \infty$, $i = 1,\ldots,m$. Let L and M be integers with $0 \leq M \leq L$. Then, for any*

$$v_j \in \cap_{k=1}^{m} H_{p_k}^{L-j}(\Omega), \ j = 0,1,\ldots,M,$$

there exists a function V(t,x) such that

$$V(t,x) \in \cap_{j=0}^{L} C^j([0,1]; \cap_{k=1}^{m} H_{p_k}^{L-j}(\Omega)), \ (\partial_t^j V)(0,x) = v_j(x), \ 0 \leq j \leq M,$$

$$\sup_{0 \leq t \leq 1} \|\overline{D}^K V(t,\cdot)\|_{p_i} \leq C(p_i,K)\Sigma_{j=0}^{M} \|v_j\|_{p_i,K-j}, \quad 1 \leq i \leq m, \ M \leq K \leq L.$$

Let u_0, u_1, f and g satisfy the N-1 th order compatibility condition and u_j, $j = 2,\ldots,N+1$, be functions defined by (5.4). If

$$u_0 \in B^{N+1}(\overline{\Omega}) \cap H_{q(n)}^{N+1}(\Omega), \ u_1 \in B^N(\overline{\Omega}) \cap H_{q(n)}^{N}(\Omega),$$

$$f \in B^{N-1}([0,\infty)\times\overline{\Omega}) \cap E_{q(n)}^{N-1}, \ \|u_0\|_{\infty,N+1} + \|u_1\|_{\infty,N} + \|\overline{D}^{N-1} f(0,\cdot)\|_\infty \leq 1,$$

then we have by (5.5) and Lemma 2.3.1

(5.8) $\|u_k\|_{r,L} \leq C(r,L)[\ \|u_0\|_{r,L+k} + \|u_1\|_{r,L+k-1} + \|\bar{D}^{L+k-2}f(0,\cdot)\|_r]$

for any integers k and L with $k+L \leq N+1$ and $k \geq 2$ and for $r = q(n)$ and ∞.

Lemma 5.3. Let u_0, u_1, f and g satisfy the N-1 th order compatibility condition and (5.7). Then, there exists a function v(t,x) such that

(a) $v(t,x) \in \cap_{j=0}^{N+1} C^j([0,\infty);\ \mathcal{B}^{N+1-j}(\bar{\Omega}) \cap H_{q(n)}^{N+1-j}(\Omega))$,

(b) $\partial_t^j v(0,x) = u_j(x)$, $j = 0,1,\ldots,N+1$,

(c) $\partial_t^j[\Phi_b(\tilde{\Lambda}v)]\big|_{t=0} = (\partial_t^j g)(0,x)$, $x \in \partial\Omega$, $j = 0,1,\ldots,N-1$,

(d) $\partial_t^j[\Phi(\Lambda v)]\big|_{t=0} = (\partial_t^j f)(0,x)$, $j = 0,1,\ldots,N-1$,

(e) $|v|_{k,N+1,d(n)r(n)} \leq c_2[\ \|u_0\|_{k,N+1} + \|u_1\|_{k,N} + \|\bar{D}^{N-1}f(0,\cdot)\|_k]$

for $k = q(n)$ and ∞. Here, c_2 is some positive constant depending only on n, Ω, N, α_{ij}, β and γ.

Proof. Let V be a function stated in Lemma 5.2 with $L = M = N+1$, $m = 2$, $p_1 = q(n)$, $p_2 = \infty$ and $v_j = u_j$, $j = 0,\ldots,N+1$. Choose $\rho(t) \in C_0^\infty(\mathbb{R}^1)$ so that $\rho(t) = 1$ if $|t| \leq 1/4$ and $= 0$ if $|t| \geq 1/2$ and put $v(t,x) = \rho(t)V(t,x)$. By (5.8), Lemma 5.2, (5.3), (5.4) and the definition of the compatibility condition, we have that v(t,x) satisfies the required properties.

Q.E.D.

From now, v(t,x) always refers to the function stated in Lemma 5.3. If u is a solution of (1.1), putting $u = v + w$, w satisfies the equation:

$$\begin{aligned} &\Box w + \Psi(\Lambda w) = F(t,x) && \text{in } [0,\infty)\times\Omega, \\ (5.9)\qquad &\frac{\partial w}{\partial \nu} + \Psi_b(\tilde{\Lambda} w) = G(t,x) && \text{on } [0,\infty)\times\partial\Omega, \\ &w(0,x) = (\partial_t w)(0,x) = 0 && \text{in } \Omega, \end{aligned}$$

where

$$\begin{aligned} &\Psi(\Lambda w) = \Phi(\Lambda v + \Lambda w) - \Phi(\Lambda v), \\ (5.10)\qquad &\Psi_b(\tilde{\Lambda} w) = \Phi_b(D_x^1 v + D_x^1 w) - \Phi_b(D_x^1 v) + \gamma(v + w) - \gamma(v), \\ &F = f - \Phi(\Lambda v), \quad G = g - \Phi_b(\tilde{\Lambda} v), \end{aligned}$$

(cf. (5.1)). Conversely, if w is a solution of (5.9), then $u = v + w$ is a solution of (1.1). Thus, we solve (5.9) below. In particular, it follows from Lemma 5.3 and (5.10) that

$$(5.11)\qquad (\partial_t^j F)(0,x) = (\partial_t^j G)(0,x)\Big|_{x\in\partial\Omega} = 0, \quad j = 0,1,\dots,N-1.$$

6. Iteration Scheme

In this section, we describe our iteration scheme for solving (5.9). The scheme constructed below is called the Nash-Moser-Hörmander scheme (see Klainerman [3]). Let θ be a fixed constant ≥ 1 and put $\theta_j = \theta^j$. Let $S(\tau)$ be a smoothing operator defined in paragraph 2.1 of section II and put $S_j u = S(\theta_j)u$. By Lemma 2.1.1 we have the following properties of smoothing operator S_j: if $u \in \mathbb{E}_p^{L,k} \cap C^L([0,\infty)\times\bar{\Omega})$ $(k \geq 0)$, then

(Sm.1) $\quad S_j u \in \mathbb{E}_p^{\infty,k}$,

(Sm.2) $\quad |(I-S_j)u|_{p,M,k} \leq C(p,M,L,k)\theta_j^{M-L}|u|_{p,L,k}$,

$$\text{(Sm.3)} \qquad |S_j u|_{p,M,k} \leqq C(p,M,L,k)\theta_j^{M-L}|u|_{p,L,k}, \qquad M \geqq L,$$

$$\text{(Sm.4)} \qquad |S_j u|_{p,M,k} \leqq C(p,M,k)|u|_{p,M,k}, \qquad M \leqq L,$$

where all constants are independent of j and u.

Let us fix w_0 by requiring that

$$\text{(6.1)} \qquad \begin{cases} \Box w_0 = F \quad \text{in } [0,\infty)\times\Omega, \quad \dfrac{\partial w_0}{\partial \nu} = G \quad \text{in } [0,\infty)\times\partial\Omega, \\ w_0(0,x) = (\partial_t w_0)(0,x) = 0 \quad \text{in } \Omega, \end{cases}$$

where F and G are the same as in (5.10). We shall define w_j, $j \geqq 1$, successively

$$\text{(6.2)} \qquad w_j = w_{j-1} + \dot{w}_{j-1} = \Sigma_{k=0}^{j-1}\dot{w}_k + w_0.$$

For this, we prepare some notations. Let us fix $\phi(x) \in C_0^\infty(\mathbb{R}^n)$ below by requiring that $\phi(x) = 1$ if $|x| \leqq R-1$ and $= 0$ if $|x| \geqq R$ (cf. Notation in section I). Put

$$\text{(6.3)} \qquad \begin{aligned} L_j u &= \Box u + (d\mathfrak{F})(S_j \Lambda w_j)\Lambda u, \\ B_j u &= \frac{\partial u}{\partial \nu} + (d\mathfrak{F}_b)(S_j D_x^1 w_j)D_x^1 u + \gamma'(S_j(\phi w_j))u, \\ e'_{b,j} &= (d\mathfrak{F}_b)(D_x^1 w_j)D_x^1\dot{w}_j - (d\mathfrak{F}_b)(S_j D_x^1 w_j)D_x^1\dot{w}_j + \\ &\qquad \phi[\gamma'(\phi w_j)\dot{w}_j - \gamma'(S_j(\phi w_j))\dot{w}_j], \\ e''_{b,j} &= \mathfrak{F}_b(D_x^1 w_{j+1}) - \mathfrak{F}_b(D_x^1 w_j) - (d\mathfrak{F}_b)(D_x^1 w_j)D_x^1\dot{w}_j + \\ &\qquad \phi[\gamma(w_{j+1}) - \gamma(w_j) - \gamma'(w_j)\dot{w}_j], \\ e'_j &= (d\mathfrak{F})(\Lambda w_j)\Lambda\dot{w}_j - (d\mathfrak{F})(S_j\Lambda w_j)\Lambda\dot{w}_j, \end{aligned}$$

$$e''_j = \tilde{\Phi}(\Lambda w_{j+1}) - \tilde{\Phi}(\Lambda w_j) - (d\tilde{\Phi})(\Lambda w_j)\Lambda\dot{w}_j,$$

$$e_j = e'_j + e''_j, \quad e_{b,j} = e'_{b,j} + e''_{b,j},$$

$$E_0 = E_{b,0} = 0, \quad E_j = \Sigma_{k=0}^{j-1} e_k, \quad E_{b,j} = \Sigma_{k=0}^{j-1} e_{b,k}, \qquad j \geq 1,$$

$$f_0 = -S_0\Psi(\Lambda w_0), \quad g_0 = -S_0[\phi\Psi_b(\tilde{\Lambda} w_0)],$$

$$f_j = -(S_j - S_{j-1})\Psi(\Lambda w_0) - (S_j - S_{j-1})E_{j-1} - S_j e_{j-1}, \qquad j \geq 1,$$

$$g_j = -(S_j - S_{j-1})[\phi\Psi_b(\tilde{\Lambda} w_0)] - (S_j - S_{j-1})E_{b,j-1} - S_j e_{j-1}, \qquad j \geq 1.$$

We define successively $\dot{w}_j$, $j \geq 0$, as solutions of the following equations:

$$\text{(6.4)} \quad \begin{aligned} &L_j\dot{w}_j = f_j \quad \text{in } [0,\infty)\times\Omega, \quad B_j\dot{w}_j = g_j \quad \text{on } [0,\infty)\times\partial\Omega, \\ &\dot{w}_j(0,x) = (\partial_t\dot{w}_j)(0,x) = 0 \quad \text{in } \Omega. \end{aligned}$$

Noting that

$$d\Psi = d\tilde{\Phi}, \; d\Psi_b = (d\tilde{\Phi}_b, \gamma'), \; \Psi(\Lambda w_{j+1}) - \Psi(\Lambda w_j) = \tilde{\Phi}(\Lambda w_{j+1}) - \tilde{\Phi}(\Lambda w_j),$$

$$\Psi_b(\Lambda w_{j+1}) - \Psi_b(\Lambda w_j) = \tilde{\Phi}_b(D_x^1 w_{j+1}) - \tilde{\Phi}_b(D_x^1 w_j) + \gamma(w_{j+1}) - \gamma(w_j),$$

$$\phi = 1 \text{ on } \partial\Omega,$$

by (6.2), (6.3) and (6.4) we have

$$\text{(6.5)} \quad \begin{aligned} &\Box w_{j+1} + \Psi(\Lambda w_{j+1}) = F + (I-S_j)\Psi(\Lambda w_0) + (I-S_j)E_j + e_j \quad \text{in } [0,\infty)\times\Omega, \\ &\frac{\partial w_{j+1}}{\partial\nu} + \Psi_b(\tilde{\Lambda} w_{j+1}) = \\ &\qquad G + (I-S_j)(\phi\Psi_b(\tilde{\Lambda} w_0)) + (I-S_j)E_{b,j} + e_{b,j} \quad \text{on } [0,\infty)\times\partial\Omega, \\ &w_{j+1}(0,x) = (\partial_t w_{j+1})(0,x) = 0 \qquad \text{in } \Omega. \end{aligned}$$

7. Estimates of Non-Linear Terms.

In order to evaluate e_j, $e_{b,j}$ and coefficients of linear operators L_j and B_j, we give some calculus lemmas in this section. Let D be a domain with compact and C^∞ boundary and $H(t,x,y)$, $y = (y_1,\ldots,y_s)$, a function in $B^\infty([0,\infty)\times\bar{D}\times\{y \in \mathbb{R}^s ; |y| \leq 1\})$. Assume that

$$H(t,x,y) = O(|y|^{r(n)+1}) \quad \text{near } y = 0. \tag{7.1}$$

Put

$$\begin{aligned} e'(H;U,V,W) &= (dH)(U)W - (dH)(V)W, \\ e''(H;U,V) &= H(U+V) - H(U) - (dH)(U)V, \end{aligned} \tag{7.2}$$

where $U = (U_1(t,x),\ldots,U_s(t,x))$, $V = (V_1(t,x),\ldots,V_s(t,x))$ and $W = (W_1(t,x),\ldots,W_s(t,x))$ are vector-valued functions.

Lemma 7.1. (a) Assume that $n \geq 6$. Then, we have

$$|(d^2H)(U)|_{D,\infty,L,0} \leq C(L,H,D)[\ 1 + |U|_{D,\infty,L,0}]$$

for any $U \in E^L_\infty(D)$ such that $|U|_{D,\infty,0,0} \leq 1$.

(b) Assume that $3 \leq n \leq 5$. Then, we have

$$|(d^2H)(U)|_{D,p,L,d(n)} \leq C(H,L,D,p)[|U|_{D,p,L,d(n)} + |U|_{D,\infty,L,0}]$$

for any $U \in E^L_\infty(D) \cap E^{L,d(n)}_p(D)$ such that $|U|_{D,\infty,0,0}$, $|U|_{D,p,0,d(n)} \leq 1$.

Proof. By (7.1) we have that $(d^2H)(y) = O(|y|^{r(n)-1})$ near $y = 0$. Since $r(n) = 1$ if $n \geq 6$ and $= 2$ if $3 \leq n \leq 5$, the lemma follows immediately from Lemmas 2.3.2 and 2.3.3.

Q.E.D.

By Taylor expansion, we can write

$$
\begin{aligned}
&e'(H;U,V,W) = \int_0^1 (d^2H)(V+a(U-V))da(U-V,W), \\
(7.3)\quad & \\
&e''(H;U,V) = \int_0^1 (1-a)(d^2H)(U+aV)da(V,V).
\end{aligned}
$$

Applying Lemmas 2.3.4 and 7.1 to (7.3), we have the following immediately.

Lemma 7.2. Let p and q be extended real numbers such that $q^{-1} = r(n)p^{-1} + 2^{-1}$. Then, the following estimates hold:

(a) $|e'(H;U,V,W)|_{D,q,L,d(n)r(n)} \leq C(H,L,D,p,q)[|U-V|_{D,2,L,0}|W|_{D,p,0,d(n)}$

$+ |U-V|_{D,2,0,0}|W|_{D,p,L,d(n)} + (1+ |U|_{D,p/(r(n)-1),L,(r(n)-1)d(n)} +$

$|V|_{D,p/(r(n)-1),L,(r(n)-1)d(n)} + |U|_{D,\infty,L,0} + |V|_{D,\infty,L,0})\times$

$\times|U-V|_{D,2,0,0}|W|_{D,p,0,d(n)}]$

for any U, V and W such that

$U,\ V \in E_2^L(D) \cap E_{p/(r(n)-1)}^{L,(r(n)-1)d(n)}(D) \cap E_\infty^L(D),\ W \in E_p^{L,d(n)}(D),$

$|U|_{D,\infty,0,0},\ |V|_{D,\infty,0,0} \leq 1/2,$

$|U|_{D,p/(r(n)-1),0,(r(n)-1)d(n)},\ |V|_{D,p/(r(n)-1),0,(r(n)-1)d(n)} \leq 1/2,$

(b) $|e''(H;U,V)|_{D,q,L,d(n)r(n)} \leq C(H,L,D,p,q)[|V|_{D,2,L,0}|V|_{D,p,0,d(n)} +$

$|V|_{D,2,0,0}|V|_{D,p,L,d(n)} + (1 + |U|_{D,p/(r(n)-1),L,(r(n)-1)d(n)} +$

$|V|_{D,p/(r(n)-1),L,(r(n)-1)d(n)} + |U|_{D,\infty,L,0} + |V|_{D,\infty,L,0})\times$

$\times|V|_{D,2,0,0}|V|_{D,p,0,d(n)}]$

for any U *and* V *such that*

$$U \in E^{L,(r(n)-1)d(n)}_{p/(r(n)-1)}(D) \cap E^{L}_{\infty}(D),$$

$$V \in E^{L}_{2}(D) \cap E^{L,(r(n)-1)d(n)}_{p/(r(n)-1)}(D) \cap E^{L}_{\infty}(D),$$

$$|U|_{D,\infty,0,0},\ |V|_{D,\infty,0,0} \leq 1/2,$$

$$|U|_{D,p/(r(n)-1),0,(r(n)-1)d(n)},\ |V|_{D,p/(r(n)-1),0,(r(n)-1)d(n)} \leq 1/2.$$

Here for notational convenience we have put $\infty^{-1} = 0$ *and* $0^{-1} = \infty$.

Next, we give a lemma to evaluate the coefficients of L_j and B_j. For this, we begin with

Lemma 7.3. *The following estimates hold.*

(a) $|(dH)(U)|_{D,\infty,L,d(n)r(n)} \leq C(L,H,D)|U|_{D,\infty,L,d(n)}$

for any $U \in E^{L,d(n)}_{\infty}(D)$ *such that* $|U|_{D,\infty,0,d(n)} \leq 1$.

(b) $|(dH)(U)|_{D,p(n)/r(n),L,d(n)r(n)}$

$$\leq C(L,H,D)[\ |U|_{D,\infty,L,0} + |U|_{D,p(n),L,d(n)}]$$

for any $U \in E^{L}_{\infty}(D) \cap E^{L,d(n)}_{p(n)}(D)$ *with* $|U|_{D,\infty,0,0},\ |U|_{D,p(n),0,d(n)} \leq 1$.

Proof. By Taylor expansion, we can write

(7.4) $$(dH)(U) = \int_0^1 (d^2H)(aU)da\cdot U.$$

Applying Lemmas 2.3.4 and 7.1 to (7.4), we have the lemma immediately.

Q.E.D.

Put

$$A(\tau,f) = ((d\tilde{\Phi})(\Lambda v+S(\tau)\Lambda f),\ (d\tilde{\Phi}_b)(D_x^1 v+S(\tau)D_x^1 f),\ \gamma'(v+S(\tau)(\phi f)))$$

(cf. (6.3)). By Lemma 7.3 we have

Lemma 7.4. Assume that Assumption 1.1- 2° is satisfied. Let c_0 and $c_{1,N}$ be the same as in Theorem 3.1 with $L = N$. Assume that

$$|v|_{\infty,N+1,d(n)},\ |v|_{p(n),N+1,d(n)} \leq 1/2.$$

Then, the following assertions are valid.

[I] There exists a positive constant c_3 depending essentially only on c_0, $c_{1,N}$, α_{ij}, β and γ such that if

$$|\Lambda v|_{\infty,2,d(n)} + |v|^R_{\infty,2,d(n)} \leq c_3,\ |\Lambda v|_{p(n),1,d(n)} + |v|^R_{p(n),1,d(n)} \leq c_3$$

then for any function f such that

$$\Lambda f \in \mathbb{E}_\infty^{2,d(n)} \cap \mathbb{E}_{p(n)}^{1,d(n)},\quad f \in \mathbb{E}_{\infty,R}^{2,d(n)} \cap \mathbb{E}_{p(n),R}^{1,d(n)},$$

$$|S(\tau)\Lambda f|_{\infty,2,d(n)} + |S(\tau)f|_{\infty,2,d(n)} \leq c_3,$$

$$|S(\tau)\Lambda f|_{p(n),1,d(n)} + |S(\tau)f|_{p(n),1,d(n)} \leq c_3,$$

the following four assertions are valid:

(a) $$\Sigma_{ij=1}^n (\delta_{ij} + \alpha'_{ij}(D_x^1 v + S(\tau)D_x^1 f))\xi_i\xi_j \geq \frac{1}{2}|\xi|^2,$$

for all $\xi = (\xi_1,\dots,\xi_n) \in \mathbb{R}^n$, α'_{ij} being the same as in (5.1),

(b) $$|A(\tau,f)|_{\infty,0,0} \leq \min((8c_0^2)^{-1},\ c_{1,N}),$$

(c) $$|A(\tau,f)|_{\infty,2,d(n)r(n)} \leq 1,$$

(d) $$|A(\tau,f)|_{p(n)/r(n),1,d(n)r(n)} \leq 1.$$

[II] $|A(\tau,f)|_{\infty,L,d(n)r(n)} \leqq C(L)[\,1 + |S(\tau)\Lambda f|_{\infty,L,d(n)} +$

$|S(\tau)(\phi f)|_{\infty,L,d(n)}],$

_if_ $0 \leqq L \leqq N-1$, $\Lambda f \in \mathbb{E}_{\infty}^{L,d(n)}$, $f \in \mathbb{E}_{\infty,R}^{L,d(n)}$,

$|S(\tau)\Lambda f|_{\infty,0,d(n)}$, $|S(\tau)(\phi f)|_{\infty,0,d(n)} \leqq 1/2.$

[III] $|A(\tau,f)|_{p(n)/r(n),L,d(n)r(n)} \leqq C(L)[\,1 + |S(\tau)\Lambda f|_{\infty,L,0} +$

$|S(\tau)\Lambda f|_{p(n),L,d(n)} + |S(\tau)(\phi f)|_{\infty,L,0} + |S(\tau)(\phi f)|_{p(n),L,d(n)}],$

_if_ $0 \leqq L \leqq N-1$, $\Lambda f \in \mathbb{E}_{\infty}^{L} \cap \mathbb{E}_{p(n)}^{L,d(n)}$, $f \in \mathbb{E}_{\infty,R}^{L} \cap \mathbb{E}_{p(n),R}^{L,d(n)}$,

$|S(\tau)\Lambda f|_{\infty,0,0}$, $|S(\tau)\Lambda f|_{p(n),0,d(n)} \leqq 1/2,$

$|S(\tau)(\phi f)|_{\infty,0,0}$, $|S(\tau)(\phi f)|_{p(n),0,d(n)} \leqq 1/2.$

Finally, we give estimates of terms: $\Psi(\Lambda w_0)$ and $\phi\Psi_b(\Lambda w_0)$. Since

$$\Psi(\Lambda w_0) = \tilde{\Phi}(\Lambda v+\Lambda w_0) - \tilde{\Phi}(\Lambda v) = \int_0^1 (d\tilde{\Phi})(\Lambda v+a\Lambda w_0)da\cdot\Lambda w_0,$$

$$\phi\Psi_b(\tilde{\Lambda} w_0) = \phi[\tilde{\Phi}_b(D_x^1 v+D_x^1 w_0) - \tilde{\Phi}_b(D_x^1 v)] + \phi[\gamma(v+w_0) - \gamma(v)] =$$

$$\phi\int_0^1 (d\tilde{\Phi}_b)(D_x^1 v+aD_x^1 w_0)da\cdot D_x^1 w_0 + \phi\int_0^1 \gamma'(v+aw_0)da\cdot w_0,$$

applying Lemmas 2.3.4 and 7.3, we have

Lemma 7.5. _Assume that Assumption_ 1.1- 2° _is satisfied_. _Let_ v _and_ w_0 _be the same as in sections_ V _and_ VI, _respectively_. _Let_ L _be a non-negative integer_ $\leqq N-1$. _If_

$|v|_{\infty,N+1,d(n)}$, $|v|_{p(n),N+1,d(n)} \leqq 1/2,$

$|\Lambda w_0|_{p(n),0,d(n)}$, $|\Lambda w_0|_{\infty,0,0}$, $|w_0|_{\infty,0,0}^{R} \leqq 1/2,$

then, the following estimates hold:

(a) $|\Psi(\Lambda w_0)|_{2,L,d(n)r(n)} \leq$

$$C(L)(|v|_{\infty,N+1,d(n)} + |\Lambda w_0|_{\infty,L,d(n)})^{r(n)}|\Lambda w_0|_{2,L,0},$$

(b) $|\Psi(\Lambda w_0)|_{q(n),L,d(n)r(n)} \leq C(L)|\Lambda w_0|_{2,L,0} \times$

$$\times(|v|_{\infty,N+1,0} + |v|_{p(n),N+1,d(n)} + |\Lambda w_0|_{\infty,L,0} + |\Lambda w_0|_{p(n),L,d(n)})^{r(n)},$$

(c) $|\phi\Psi_b(\chi w_0)|_{2,L,d(n)r(n)} \leq$

$$C(L)(|v|_{\infty,N+1,d(n)} + |\chi w_0|^R_{\infty,L,d(n)})^{r(n)}|\chi w_0|^R_{2,L,0}.$$

8. Convergence of Iteration Scheme and Proof of Theorem 1.3

In what follows, we use notations defined in sections V and VI. First, by induction we shall prove that for all $j \geq 0$ the following statements are valid:

(St.1) $\Lambda\dot{w}_j \in \mathbb{E}_2^{N'} \cap \mathcal{B}^{N'}([0,\infty)\times\bar{\Omega})$, $\dot{w}_j \in \mathbb{E}_{2,R}^{N'} \cap \mathcal{B}^{N'}([0,\infty)\times\bar{\Omega}_R)$,

(St.2) $[\Lambda\dot{w}_j]_L + [\dot{w}_j]_L^R \leq \delta\theta_j^{-N''+L}, \quad 0 \leq L \leq N'.$

Here δ is a positive constant determined later, $\dot{w}_j$, $j = 0,1,2,\dots$ are the same as in section VI and we have put

$$[\Lambda u]_L = |\Lambda u|_{2,L,0} + |\Lambda u|_{p(n),L,d(n)} + |\Lambda u|_{\infty,L,d(n)},$$

$$[u]_L^R = |u|^R_{2,L,0} + |u|^R_{p(n),L,d(n)} + |u|^R_{\infty,L,d(n)}.$$

Let a be a non-negative integer and assume that

(A.1) $\dot{w}_0,\dots,\dot{w}_a$ are already given and the statements (St.1) and (St.2) are valid for $\dot{w}_0, \dots, \dot{w}_a$,

(A.2) w_0, being a solution of the equation (6.1), satisfies the conditions:

$$\Lambda w_0 \in \mathbb{E}_2^{N'} \cap \mathcal{B}^{N'}([0,\infty)\times\bar{\Omega}),\ w_0 \in \mathbb{E}_{2,R}^{N'} \cap \mathcal{B}^{N'}([0,\infty)\times\bar{\Omega}_R),\ [\Lambda w_0]_{N'} + [w_0]_{N'}^R \leqq \delta,$$

(A.3) $|v|_{\infty,N+1,d(n)r(n)} + |v|_{q(n),N+1,d(n)r(n)} \leqq \delta.$

In particular, from (A.3) it follows that

(A.3)' $|v|_{p,N+1,d(n)r(n)} \leqq \delta$ for all p with $q(n) \leqq p \leqq \infty$.

To prove (St.1) and (St.2), we need several lemmas. In the same way as in Klainerman [3, §7] and Shibata [8, §5], these lemmas can be proved by using Lemma 2.2.1, the properties of smoothing operators (Sm.1)-(Sm.4) in section VI, Lemmas 7.2, 7.4 and 7.5, and the fact that $\theta_j = \theta^j$, so we may omit their proofs.

Lemma 8.1. Assume that the assumptions (A.1) and (A.2) are satisfied. Put

$$w_j = \Sigma_{k=0}^{j-1} \dot{w}_k + w_0, \qquad j = 1,2,\dots,a+1.$$

Then, the following assertions are valid.

(a) $\Lambda w_j \in \mathbb{E}_2^{N'} \cap \mathcal{B}^{N'}([0,\infty)\times\bar{\Omega}),\ w_j \in \mathbb{E}_{2,R}^{N'} \cap \mathcal{B}^{N'}([0,\infty)\times\bar{\Omega}_R)$,

(b) $[\Lambda w_j]_L + [w_j]_L^R \leqq c_4\delta$, $\quad -N''+L \leqq -1$,

(c) $[\Lambda w_j]_L + [w_j]_L^R \leqq c_5\delta\theta_j^{-N''+L}$, $\quad -N''+L \geqq 1,\ 0 \leqq L \leqq N'$,

(d) $[S_j\Lambda w_j]_L + [S_j(\phi w_j)]_L^R \leqq C(L)\delta\theta_j^{-N''+L}$, $\quad L \geqq N'+1$,

(e) $[(I-S_j)\Lambda w_j]_L + [(I-S_j)(\phi w_j)]_L^R \leq c_6 \delta \theta_j^{-N''+L}$, $\quad 0 \leq L \leq N'$,

<u>where</u> c_4, c_5 <u>and</u> c_6 <u>are some positive constants depending only on</u> θ <u>and constants appearing in</u> (Sm.2)-(Sm.4) <u>of section</u> VI.

By Lemma 8.1 and (Sm.4) in section VI, we have

$$[S_j \Lambda w_j]_L + [S_j(\phi w_j)]_L \leq c_7 \delta, \qquad -N''+L \leq -1,$$

for some positive constant c_7. Assume that

(A.4) $\delta \leq 1/4$, $c_7\delta \leq 1/4$, $c_4\delta \leq 1/4$, $c_7\delta \leq c_3/2$, $c_4\delta \leq c_3$,

where c_3 and c_4 are the same as in Lemmas 7.4 and 8.1, respectively.

Lemma 8.2. <u>Assume that Assumption</u> 1.1- 2° <u>and assumptions</u> (A.1)-(A.4) <u>are satisfied. Then, the following assertions are valid for any</u> $j = 0,1,\ldots,a$.

(a) $e_j \in \mathbb{E}^{N'}_{q(n)} \cap \mathcal{B}^{N'}([0,\infty)\times\bar{\Omega})$, $e_{b,j} \in \mathbb{E}^{N'}_2 \cap \mathcal{B}^{N'}([0,\infty)\times\bar{\Omega}_R)$.

(b) $|e_j|_{2,L,d(n)r(n)} + |e_j|_{q(n),L,d(n)r(n)} \leq c_8\delta^2\theta_j^{-2N''+L}$, $\quad 0 \leq L \leq N'$.

(c) $|e_{b,j}|_{2,L,d(n)r(n)} \leq c_8\delta^2\theta_j^{-2N''+L}$, $\quad 0 \leq L \leq N'$.

<u>Here</u> c_8 <u>is some positive constant independent of</u> j <u>and</u> L.

Lemma 8.3. <u>Assume that Assumption</u> 1.1- 2° <u>and assumptions</u> (A.1)-(A.4) <u>are satisfied. Then, the following assertions are valid.</u>

(a) $\Psi(\Lambda w_0) \in \mathbb{E}^{N'}_{q(n)} \cap \mathcal{B}^{N'}([0,\infty)\times\bar{\Omega})$, $\phi\Psi_b(\lambda w_0) \in \mathbb{E}^{N'}_2 \cap \mathcal{B}^{N'}([0,\infty)\times\bar{\Omega}_R)$.

(b) $|\Psi(\Lambda w_0)|_{2,N',d(n)r(n)} + |\Psi(\Lambda w_0)|_{q(n),N',d(n)r(n)} +$

$$|\phi\Psi_b(\lambda w_0)|_{2,N',d(n)r(n)} \leq c_9\delta^2.$$

(c) $|(I-S_j)\Psi(\Lambda w_0)|_{2,L,d(n)r(n)} + |(I-S_j)\Psi(\Lambda w_0)|_{q(n),L,d(n)r(n)} +$

$$|(I-S_j)(\phi\Psi_b(\tilde{\Lambda}w_0))|_{2,L,d(n)r(n)} \leqq c_{10}\delta^2\theta_j^{-2N''+L}, \quad 0 \leq L \leq N', \ \forall j \geqq 0.$$

(d) $|(S_{a+1}-S_a)\Psi(\Lambda w_0)|_{2,L,d(n)r(n)} + |(S_{a+1}-S_a)\Psi(\Lambda w_0)|_{q(n),L,d(n)r(n)} +$

$$|(S_{a+1}-S_a)(\phi\Psi_b(\tilde{\Lambda}w_0))|_{2,L,d(n)r(n)} \leqq C(L)\delta^2\theta_{a+1}^{-2N''+L}, \qquad \forall L \geqq 0.$$

(e) $|S_0\Psi(\Lambda w_0)|_{2,L,d(n)r(n)} + |S_0\Psi(\Lambda w_0)|_{q(n),L,d(n)r(n)} +$

$$|S_0(\phi\Psi_b(\Lambda w_0))|_{2,L,d(n)r(n)} \leqq C(L)\delta^2\theta_0^{-2N''+L}, \qquad \forall L \geqq 0.$$

Here c_9 and c_{10} are some positive constants.

Lemma 8.4. Assume that Assumption 1.1- 2° and assumptions (A.1)-(A.4) are satisfied. Put

$$E_0 = E_{b,0} = 0, \ E_a = \Sigma_{j=0}^{a-1} e_j, \ E_{b,a} = \Sigma_{j=0}^{a-1} e_{b,a} \quad (\text{if } a \geqq 1).$$

Then, the following assertions are valid.

(a) $E_a \in \mathbb{E}_{q(n)}^{N'} \cap \mathcal{B}^{N'}([0,\infty)\times\bar{\Omega}), \ E_{b,a} \in \mathbb{E}_2^{N'} \cap \mathcal{B}^{N'}([0,\infty)\times\bar{\Omega}_R).$

(b) $|E_a|_{2,L,d(n)r(n)} + |E_a|_{q(n),L,d(n)r(n)} +$

$$|E_{b,a}|_{2,L,d(n)r(n)} \leqq c_{11}\delta^2, \qquad -2N''+L \leq -1.$$

(c) $|E_a|_{2,L,d(n)r(n)} + |E_a|_{q(n),L,d(n)r(n)} +$

$$|E_{b,a}|_{2,L,d(n)r(n)} \leqq c_{12}\delta^2\theta_a^{-2N''+L}, \qquad -2N''+L \geqq 1, \ 0 \leq L \leq N'.$$

(d) $|(I-S_a)E_a|_{2,L,d(n)r(n)} + |(I-S_a)E_a|_{q(n),L,d(n)r(n)} +$

$$|(I-S_a)E_{b,a}|_{2,L,d(n)r(n)} \leqq c_{13}\delta^2\theta_a^{-2N''+L}, \qquad 0 \leq L \leq N'.$$

(e) $|(S_{a+1}-S_a)E_a|_{2,L,d(n)r(n)} + |(S_{a+1}-S_a)E_a|_{q(n),L,d(n)r(n)} +$

$$|(S_{a+1}-S_a)E_{b,a}|_{2,L,d(n)r(n)} \leq C(L)\delta^2\theta_a^{-2N''+L}, \qquad \forall L \geq 0.$$

Here c_{11}, c_{12} *and* c_{13} *are some positive constants.*

Combining Lemmas 8.3 and 8.4, we have

Lemma 8.5. *Assume that Assumption* 1.1- 2° *and assumptions* (A.1)-(A.4) *are satisfied. Let* f_0, g_0, f_{a+1} *and* g_{a+1} *be the same as in* (6.3). *Then, the following assertions are valid.*

(a) $f_0, f_{a+1} \in \mathbb{E}^\infty_{q(n)} \cap \mathcal{B}^\infty([0,\infty)\times\bar{\Omega}), \quad g_0, g_{a+1} \in \mathbb{E}^\infty_2 \cap \mathcal{B}^\infty([0,\infty)\times\bar{\Omega})$

(b) $|f_{a+1}|_{2,L,d(n)r(n)} + |f_{a+1}|_{q(n),L,d(n)r(n)} +$

$$|g_{a+1}|^R_{2,L,d(n)r(n)} \leq C(L)\delta^2\theta_{a+1}^{-2N''+L}, \qquad \forall L \geq 0.$$

(c) $|f_0|_{2,L,d(n)r(n)} + |f_0|_{q(n),L,d(n)r(n)} +$

$$|g_0|^R_{2,L,d(n)r(n)} \leq C(L)\delta^2\theta_0^{-2N''+L}, \qquad \forall L \geq 0.$$

In order to use the results obtained in sections III and IV, we have to evaluate the coefficients of the operators L_j and B_j.

Lemma 8.6. *Assume that Assumption* 1.1- 2° *and assumptions* (A.1)-(A.4) *are satisfied. Put*

$$A_j = ((d\tilde{\Phi})(S_j\Lambda w_j),\ (d\tilde{\Phi}_b)(S_jD^1_xw_j),\ \gamma'(S_j(\phi w_j))) \qquad (\text{cf. } (6.3)).$$

Then, the following assertions are valid.

(a) $\Sigma^n_{ij=1}(\delta_{ij} + \alpha'_{ij}(D^1_xv+S_jD^1_xw_j))\xi_i\xi_j \geq \frac{1}{2}|\xi|^2$

for all $\xi = (\xi_1,\dots,\xi_n) \in \mathbb{R}^n$, α'_{ij} *being the same as in* (5.1).

(b) $|A_j|_{\infty,0,0} \leq \min((8c_0^2)^{-1}, c_{1,N})$,

where c_0 *and* $c_{1,N}$ *are the same as in Theorem* 3.1 *with* $L = N$.

(c) $|A_j|_{\infty,2,d(n)r(n)} \leq 1$.

(d) $|A_j|_{p(n)/r(n),1,d(n)r(n)} \leq 1$.

(e) $|A_j|_{\infty,L,d(n)r(n)} \leq c_{14}$, $\qquad -N''+L \leq -1$.

(f) $|A_j|_{p(n)/r(n),L,d(n)r(n)} \leq c_{15}$, $\qquad -N''+L \leq -1$.

(g) $|A_j|_{\infty,L,d(n)r(n)} \leq C(L)(1+\delta\theta_j^{-N''+L})$ $\qquad -N''+L \geq 1,\ 0 \leq L \leq N-1$.

(h) $|A_j|_{p(n)/r(n),L,d(n)r(n)} \leq C(L)(1+\delta\theta_j^{-N''+L})$, $-N''+L \geq 1,\ 0 \leq L \leq N-1$.

Here c_{14} *and* c_{15} *are some positive constants.*

Now, we give estimates of $\dot{w}_{a+1}$ and $\dot{w}_0$. Note that

$$L_j u = \Box u + (d\tilde{\Phi})(S_j \Lambda w_j)\Lambda u =$$

$$\partial_t^2 u - \Sigma_{k,m=1}^n \partial_k((\delta_{km} + \alpha'_{km}(S_j D_x^1 w_j))\partial_m u) + \Sigma_{m=0}^n (\frac{\partial\beta}{\partial\lambda_m})(S_j D^1 w_j)\partial_m u,$$

$$B_j u = \frac{\partial u}{\partial\nu} + (d\tilde{\Phi}_b)(S_j D_x^1 w_j) D_x^1 u + \gamma'(S_j(\phi w_j))u =$$

$$\Sigma_{k,m=1}^n \nu_k(\delta_{km} + \alpha'_{km}(S_j D_x^1 w_j))\partial_m u + \gamma'(S_j(\phi w_j))u,$$

α'_{ij} being the same as in (5.1). By virture of Lemma 8.6, we can apply Theorems 3.1 and 4.1 to (6.4). First, by Theorem 3.1 with $L = N$, Lemmas 8.5 and 8.6 we have

(8.1) $\quad |D^1\dot{w}_{a+1}|_{2,L,0} \leq c_{16}\delta^2\theta_{a+1}^{-2N''+L+2} \qquad 0 \leq L \leq N-1.$

By Lemma 2.4.2 and (8.1) we have

$$(8.2)\quad \begin{aligned} &|\Lambda\dot{w}_{a+1}|_{2,L,0} \leqq C(L)\delta^2\theta_{a+1}^{-2N''+L+3}, && 0 \leqq L \leqq N-2,\\ &|\dot{w}_{a+1}|^R_{2,L,0} \leqq C(L)\delta^2\theta_{a+1}^{-2N''+L+2}, && 0 \leqq L \leqq N-1. \end{aligned}$$

In particular, if

$$(8.3)\quad (\max_{0\le L\le N'} C(L)) \leqq 1/4, \qquad C(L) \text{ being the same as in (8.2)},$$

by the fact that $N'' \geqq 3$ and (8.2) we have

$$(8.4)\quad \begin{aligned} &|\Lambda\dot{w}_{a+1}|_{2,L,0} \leqq 4^{-1}\delta\theta_{a+1}^{-N''+L}, && 0 \leqq L \leqq N',\\ &|\dot{w}_{a+1}|^R_{2,L,0} \leqq 4^{-1}\delta\theta_{a+1}^{-N''+L}, && 0 \leqq L \leqq N'. \end{aligned}$$

Next, we apply Theorem 4.1. Since $N'+[n/p(n)]+2 = N-1-(2n+5)$, applying Theorem 4.1 with $L = N-1$, by Sobolev's inequality, Lemmas 8.5 and 8.6 and (8.2) we have

$$(8.5)\quad \begin{aligned} &|D^1\dot{w}_{a+1}|_{\infty,L,d(n)} + |D^1\dot{w}_{a+1}|_{p(n),L,d(n)}\\ &\leqq C|D^1\dot{w}_{a+1}|_{p(n),L+[n/p(n)]+1,d(n)}\\ &\leqq c_{17}\delta^2\theta_{a+1}^{-2N''+L+[n/p(n)]+2n+7}, \qquad 0 \leqq L \leqq N'+1, \end{aligned}$$

$$(8.6)\quad \begin{aligned} &|\dot{w}_{a+1}|^R_{\infty,0,d(n)} + |\dot{w}_{a+1}|^R_{p(n),0,d(n)}\\ &\leqq C|\bar{D}^{-1}\dot{w}_{a+1}|^R_{p(n),[n/p(n)],d(n)} \leqq c_{18}\delta^2\theta_{a+1}^{-2N''+2n+6}. \end{aligned}$$

If we choose δ so small that

$$(8.7)\quad \delta c_{17} \leqq 1/4, \quad \delta c_{18} \leqq 1/4,$$

then, by the fact that $N'' \geqq 2n+[n/p(n)]+8$, (8.5) and (8.6) we have

$$(8.8)\quad \begin{aligned} &|\Lambda\dot{w}_{a+1}|_{p(n),L,d(n)} + |\Lambda\dot{w}_{a+1}|_{\infty,L,d(n)} \leqq 4^{-1}\delta\theta_{a+1}^{-N''+L} && 0 \leqq L \leqq N', \\ &|\dot{w}_{a+1}|^R_{p(n),L,d(n)} + |\dot{w}_{a+1}|^R_{\infty,L,d(n)} \leqq 4^{-1}\delta\theta_{a+1}^{-N''+L}, && 0 \leqq L \leqq N'. \end{aligned}$$

Combining (8.4) and (8.8), we have

$$(8.9)\qquad [\Lambda\dot{w}_{a+1}]_L + [\dot{w}_{a+1}]^R_L \leqq \delta\theta_{a+1}^{-N''+L}, \qquad 0 \leqq L \leqq N',$$

if δ satisfies (A.4), (8.3) and (8.7). In the same way, we have

$$(8.10)\qquad [\Lambda\dot{w}_0]_L + [\dot{w}_0]^R_L \leqq \delta\theta_0^{-N''+L} \qquad 0 \leqq L \leqq N'.$$

Thus, by induction we obtain that the statements (St.1) and (St.2) are valid for all $j \geqq 0$ under the assumptions (A.2) and (A.3).

Now, we give a condition for data in order that (A.2) and (A.3) are valid. Since w_0 is a solution of (6.1), by (4.6) and (4.7), Theorem 3.1 with $A = 0$ and Lemma 2.4.2 we have

$$\begin{aligned} &|\Lambda w_0|_{2,N',0} + |w_0|^R_{2,N',0} \\ &\quad \leqq c_{19}[|F|_{2,N'+1,d(n)r(n)} + |G|^R_{2,N'+3,d(n)r(n)}], \end{aligned}$$

$$(8.11)\quad \begin{aligned} &|\Lambda w_0|_{\infty,N',d(n)} + |w_0|^R_{\infty,N',d(n)} + \\ &\qquad\qquad |\Lambda w_0|_{p(n),N',d(n)} + |w_0|^R_{p(n),N',d(n)} \\ &\quad \leqq C[|D^1 w_0|_{p(n),N'+[n/p(n)]+2,d(n)} + |\bar{D}^1 w_0|^R_{p(n),N'+[n/p(n)],d(n)}] \\ &\quad \leqq c_{20}[|F|_{q(n),N'+[n/p(n)]+2n+4,d(n)r(n)} + \\ &\qquad\qquad |G|^R_{2,N'+[n/p(n)]+2n+6,d(n)r(n)}]. \end{aligned}$$

By (5.10), the fact that $N = N'+[n/p(n)]+8$ and Lemma 2.3.3 we have

(8.12.a) $|F|_{2,N'+1,d(n)r(n)} + |G|^R_{2,N'+3,d(n)r(n)}$

$\leqq |f|_{2,N'+1,d(n)r(n)} + |g|^R_{2,N'+3,d(n)r(n)} + C|v|_{2,N'+4,d(n)r(n)}$,

(8.12.b) $|F|_{q(n),N'+[n/p(n)]+2n+4,d(n)r(n)}$

$\leqq |f|_{q(n),N-4,d(n)r(n)} + C|v|_{q(n),N'+3,d(n)r(n)}$,

(8.12.c) $|G|^R_{2,N'+[n/p(n)]+2n+6,d(n)r(n)}$

$\leqq |g|^R_{2,N-2,d(n)r(n)} + C|v|_{2,N,d(n)r(n)}$

provided that $|v|_{\infty,N+1,0} \leqq 1$. Noting (A.3)' and combining (8.10), (8.11) and (8.12), we have that there exists a constant $c_{21} > 0$ such that if

(8.13)
$$|f|_{q(n),N-4,d(n)r(n)} + |f|_{2,N'+1,d(n)r(n)} + |g|^R_{2,N-2,d(n)r(n)} + |v|_{\infty,N+1,d(n)r(n)} + |v|_{q(n),N+1,d(n)r(n)} \leqq c_{21}\delta,$$

then (A.2) is valid. Therefore, by Lemma 5.3 we have that there exists a positive constant δ_1 such that if (1.5) holds then (A.2) and (A.3) are valid. If we put

$$w = \Sigma^{\infty}_{j=0} \hat{w}_j + w_0,$$

noting that $N'' \geqq \ell+1$, by (St.2) we have

$$[\Lambda w]_\ell + [w]^R_\ell \leqq (1+(\theta-1)^{-1})\delta.$$

By (6.5), Lemmas 8.2, 8.3 and 8.4 we have that w satisfies the equation (5.9), which proves the existence theorem for (1.1).

Finally, we prove the uniqueness theorem for (1.1). For this, we begin with

Lemma 8.7. Let r and T be any positive numbers with $r \geq R$ and b a large number with $b \geq 2(n+1)$. Put

$$\Gamma(r,T) = \{(t,x);\ x \in \Omega,\ |x| \leq r+b(T-t),\ 0 \leq t \leq T\}.$$

Let $a_{ij}(t,x)$, $i,j = 1,\dots,n$, $b_j(t,x)$, $j = 0,\dots,n$, $c(t,x)$ be real-valued functions in $C^1([0,\infty)\times\bar{\Omega})$ such that

$$\sup_{(t,x)\in\Gamma(r,T)} |a_{ij}(t,x)| \leq 1/2, \quad a_{ij} = a_{ji}.$$

If $u \in C^2([0,\infty)\times\bar{\Omega})$ satisfies the equations:

$$Pu = \partial_t^2 u - \Sigma_{ij=1}^n \partial_i((\delta_{ij} + a_{ij})\partial_j u) + \Sigma_{j=0}^n b_j\partial_j u = 0 \quad \text{in } \Gamma(r,T),$$

$$\Sigma_{ij=1}^n \nu_i(\delta_{ij} + a_{ij})\partial_j u + cu = 0 \quad \text{on } [0,T]\times\partial\Omega,$$

$$u(0,x) = (\partial_t u)(0,x) = 0 \quad \text{in } \Omega_{r+bT},$$

then $u = 0$ in $\Gamma(r,T)$.

Integrating $Pu\cdot\partial_t u$ over $\Gamma(r,T)$ and using the divergence theorem, we obtain Lemma 8.7 in the usual way. So, we may omit the proof.

Using Lemma 8.7, we shall prove the uniqueness theorem. Let u and v be solutions in $C^2([0,\infty)\times\bar{\Omega})$ of (1.1) for the same data such that

$$(8.14)\quad |\Lambda u|_{\infty,0,0} + |u|^R_{\infty,0,0} \leq 1,\ |\Lambda v|_{\infty,0,0} + |v|^R_{\infty,0,0} \leq 1,\ |D_x^1 u|_{\infty,0,0} \leq \delta_2,$$

where δ_2 is a constant determined later. By Taylor expansion,

$$(8.15)\quad \Sigma_{ij=1}^n \partial_i(\alpha_{ij}(D_x^1 u)\partial_j u) - \Sigma_{ij=1}^n \partial_i(\alpha_{ij}(D_x^1 v)\partial_j v)$$

$$= \Sigma_{ij=1}^n \partial_i(a_{ij}(D_x^1 u,\ D_x^1 v)\partial_j(u-v)),$$

$$\Sigma_{ij=1}^n \nu_i(x)\alpha_{ij}(D_x^1 u)\partial_j u - \Sigma_{ij=1}^n \nu_i(x)\alpha_{ij}(D_x^1 v)\partial_j v$$

$$= \Sigma_{ij=1}^n \nu_i(x)a_{ij}(D_x^1 u,\ D_x^1 v)\partial_j(u-v),$$

$$\beta(D^1 u) - \beta(D^1 v) = \Sigma_{j=0}^n b_j(D^1 u,\ D^1 v)\partial_j(u-v),$$

$$\gamma(u) - \gamma(v) = c(u,v)(u-v).$$

Here we have put

$$\begin{aligned} a_{ij} &= a_{ij}(D_x^1 u,\ D_x^1 v) \\ &= \alpha_{ij}(D_x^1 u) + \Sigma_{k=1}^n \int_0^1 (\partial_{\lambda_j}\alpha_{ik})(D_x^1 v+s(D_x^1 u-D_x^1 v))ds\cdot\partial_k v, \\ b_j &= b_j(D^1 u,\ D^1 v) = \int_0^1 (\partial_{\lambda_j}\beta)(D^1 v+s(D^1 u-D^1 v))ds, \\ c &= c(u,v) = \int_0^1 \gamma'(v+s(u-v))ds. \end{aligned} \tag{8.16}$$

Putting $w = u-v$, we have by (8.15)

$$\begin{aligned} &\partial_t^2 w - \Sigma_{ij=1}^n \partial_i((\delta_{ij} + a_{ij})\partial_j w) + \Sigma_{j=0}^n b_j\partial_j w = 0 && \text{in } [0,\infty)\times\Omega, \\ &\Sigma_{ij=1}^n \nu_i(\delta_{ij} + a_{ij})\partial_j w + cw = 0 && \text{on } [0,\infty)\times\partial\Omega, \\ &w(0,x) = (\partial_t w)(0,x) = 0 && \text{in } \Omega. \end{aligned} \tag{8.17}$$

Since $v(t,x) = u(0,x) + \int_0^t (\partial_t v)(s,x)ds$, by (8.14) we have

$$|D_x^1 v(t,x)| \leq 2\delta_2, \quad (t,x) \in [0,\delta_2]\times\overline{\Omega}. \tag{8.18}$$

Since by (8.16) and Assumption 1.1- 2° we have

$$\Sigma_{ij=1}^n |a_{ij}(D_x^1,\ D_x^1 v)| \leq c_{22}(|D_x^1 u| + |D_x^1 v|) \tag{8.19}$$

for some positive constant c_{22} depending only on α_{ij}, if we choose δ_2 so small that

(8.20) $3\delta_2 c_{22} \leqq 1/2$,

we have by (8.14), (8.18), (8.19) and (8.20)

(8.21) $\Sigma_{ij=1}^n |a_{ij}(t,x)| \leqq 1/2, \quad (t,x) \in [0,\delta_2]\times\bar{\Omega}$.

Since $u, v \in C^2([0,\infty)\times\bar{\Omega})$, we have that $a_{ij}, b_j, c \in C^1([0,\infty)\times\bar{\Omega})$. Applying Lemma 8.7 to (8.17), we have that $w = u - v = 0$ in $\Gamma(r,\delta_2)$ for any $r \geqq R$, which implies that $u = v$ in $[0,\delta_2]\times\bar{\Omega}$. Since δ_2 depends only on α_{ij} (cf. (8.19) and (8.20)), replacing 0 by δ_2 and δ_2 by $2\delta_2$ and repeating the argument, we have that $u = v$ in $[\delta_2, 2\delta_2]\times\bar{\Omega}$, which implies that $u = v$ in $[0,2\delta_2]\times\bar{\Omega}$. By repeated use of the argument, we have that $u = v$ in $[0,T]\times\bar{\Omega}$ for any $T > 0$, which implies that $u = v$ in $[0,\infty)\times\bar{\Omega}$. This completes the proof of the uniqueness theorem for (1.1).

Appendix. Proofs of (4.4) and (4.5)

In order to prove (4.4) and (4.5), we need the following two lemmas.

Lemma Ap.1 (local energy decay). Let a, b and d be real numbers with $0 < d \leqq n-1$ and $a, b \geqq R-1$. Assume that $n \geqq 3$ and Ω is non-trapping. Let u be a C^∞ solution of (4.3) for data u_0, u_1, f and g. Assume that all norms appearing below are finite. If $g = 0$, supp $u_i \subset \bar{\Omega}_a$, $i = 0,1$, and supp $f \subset [0,\infty)\times\bar{\Omega}_a$, then

$$\|\bar{D}^{L+1}u(t,\cdot)\|_{\Omega_b,2} \leqq C(L,a,b,\Omega)(1+t)^{-d}[\,\|u_0\|_{2,L+1} + \|u_1\|_{2,L} + |f|_{2,L,d}].$$

If $n \geqq 3$ and n is odd, Lemma Ap.1 follows from Morawetz [7]. If $n \geqq 4$ and n is even, Lemma Ap.1 can be proved in the same way as in

Shibata and Tsutsumi [12]. For further details, Tsutsumi [14] can be referred. The following result is well-known (see Wahl [16]).

Lemma Ap. 2. Let p be an extended real number and $q = p/(p-1)$. Let v be a solution of Cauchy problem:

$$\Box v = 0 \text{ in } [0,\infty)\times\mathbb{R}^n,\ v(0,x) = v_0(x),\ (\partial_t v)(0,x) = v_1(x) \text{ in } \mathbb{R}^n.$$

Then, the following two estimates hold.

(a) $$\|\overline{D}^L v(t,\cdot)\|_{\mathbb{R}^n,p} \leq C(1+t)^{-\frac{n-1}{2}(1-\frac{2}{p})}[\,\|v_0\|_{\mathbb{R}^n,q,L+n+1} + \|v_1\|_{\mathbb{R}^n,q,L+n}],$$

$$t > 0,\ p > \frac{2(n+1)}{n-1},\ v_0 \in H_q^{L+n+1}(\mathbb{R}^n),\ v_1 \in H_q^{L+n}(\mathbb{R}^n).$$

(b) $$\|\overline{D}^L D^1 v(t,\cdot)\|_{\mathbb{R}^n,p} \leq C(1+t)^{-\frac{n-1}{2}(1-\frac{2}{p})}[\,\|v_0\|_{\mathbb{R}^n,q,L+n+2} + \|v_1\|_{\mathbb{R}^n,q,L+n+1}],$$

$$t > 0,\ 2 \leq p \leq \infty,\ v_0 \in H_q^{L+n+2}(\mathbb{R}^n),\ v_1 \in H_q^{L+n+1}(\mathbb{R}^n).$$

Now, we prove (4.4) and (4.5). In the course of the proof, by $S(D) = S(t,x;D)$, $D = (u_0,u_1,f)$, we denote the solution of Neumann problem:

$$\Box u = f \text{ in } [0,\infty)\times\Omega,\ \frac{\partial u}{\partial \nu} = 0 \text{ on } [0,\infty)\times\partial\Omega,\ u\big|_{t=0} = u_0,\ \partial_t u\big|_{t=0} = u_1 \text{ in } \Omega.$$

And also, by $S_0(D_0) = S_0(t,x;D_0)$, $D_0 = (v_0,v_1,g)$, we denote the solution of Cauchy problem:

$$\Box v = g \text{ in } [0,\infty)\times\mathbb{R}^n,\ v\big|_{t=0} = v_0,\ \partial_t v\big|_{t=0} = v_1 \text{ in } \mathbb{R}^n.$$

The proof is divided into three steps.

1st step. We consider the case where $f = 0$, supp $u_i \subset \mathbb{R}^n - \Omega_{R+2}$,

$i = 0,1$. In the same way as in Shibata and Tsutsumi [11, Proof of Lemma 4.7], by using Lemmas Ap.1 and Ap.2, we have

$$\text{(Ap.1)} \qquad \|\bar{D}^{M+1}S(t,\cdot;D)\|_\infty \leq C(\alpha)(1+t)^{-\alpha}[\ \|u_0\|_{1,M+2n+3} + \|u_1\|_{1,M+2n+2}]$$

for any $t > 0$, where $M = 0$ or L, $D = (u_0,u_1,0)$ and we have put

$$\text{(Ap.2)} \qquad \alpha = \frac{n-1}{2} \text{ if } n \geq 4 \text{ and } 1-\varepsilon \text{ if } n = 3$$

for any small positive number ε.

On the other hand, by the usual energy method (cf. Theorem 3.1 with $A = g = 0$) and Lemma 2.4.2 we have

$$\text{(Ap.3)} \qquad \|\bar{D}^L D^1 u(t,\cdot)\|_2 \leq C(L)[\ \|u_0\|_{2,L+1} + \|u_1\|_{2,L}],$$

$$\text{(Ap.3)'} \qquad \|\bar{D}^1 u(t,\cdot)\|^R \leq C[\ \|u_0\|_{2,1} + \|u_1\|\ ].$$

Interpolating (Ap.1) with $M = L$ and (Ap.3) and interpolating (Ap.1) with $M = 0$ and (Ap.3)', we have

$$\text{(Ap.4)} \qquad \begin{aligned} &\|\bar{D}^L D^1 u(t,\cdot)\|_p \leq C(\alpha)(1+t)^{-\alpha(1-\frac{2}{p})}[\ \|u_0\|_{q,L+2n+3} + \|u_1\|_{q,L+2n+2}], \\ &\|\bar{D}^1 u(t,\cdot)\|_p^R \leq C(\alpha)(1+t)^{-\alpha(1-\frac{2}{p})}[\ \|u_0\|_{q,2n+3} + \|u_1\|_{q,2n+2}], \end{aligned}$$

for any p with $2 \leq p \leq \infty$ and $q = p/(p-1)$.

<u>2nd step</u>. We consider the case where $u_0 = u_1 = 0$ and supp $f \subset [0,\infty)\times(\mathbb{R}^n-\Omega_{R+2})$. In this case, we can use Duhamel's principle for the mixed problem. Thus, we can write

$$\text{(Ap.5)} \qquad S(t,x;D) = \int_0^t S(t-s,x,D'(s))ds$$

where $D = (0,0,f)$ and $D'(s) = (0,f(s,\cdot),0)$. Note that

$$\int_0^t (1+(t-s))^{-\alpha(1-\frac{2}{p})}(1+s)^{-\beta}\, ds \leq C(1+t)^{-\alpha(1-\frac{2}{p})}$$

where

(Ap.6) $\beta = \alpha(1-\frac{2}{p})$ if $\alpha(1-\frac{2}{p}) > 1$ and $= 1+\kappa$ if $\alpha(1-\frac{2}{p}) \leq 1$

for any sufficiently small positive number κ. Applying (Ap.4) to (Ap.5), we have

(Ap.7)
$$\|\bar{D}^1 S(t,\cdot;D)\|_p^R \leq C(\alpha)(1+t)^{-\alpha(1-\frac{2}{p})}|f|_{q,2n+2,\beta},$$
$$\|\bar{D}^L D^1 S(t,\cdot;D)\|_p \leq C(\alpha,L)(1+t)^{-\alpha(1-\frac{2}{p})}|f|_{q,L+2n+2,\beta},$$

for any $t > 0$. In particular, combining (Ap.4) and (Ap.7) and taking $p = p(n)$, $q = q(n)$, $\alpha(1-\frac{2}{p(n)}) = d(n)$, $\beta = d(n)r(n)$, we obtain (4.4) and (4.5) when supp $u_i \subset \mathbb{R}^n - \Omega_{R+2}$, $i = 0,1$, and supp $f \subset [0,\infty)\times(\mathbb{R}^n - \Omega_{R+2})$.

3rd step. To complete the proof, we prove (4.4) and (4.5) under the assumptions:

(Ap.8) supp $u_i \subset \Omega_{R+3}$, $i = 0,1$, supp $f \subset [0,\infty)\times\Omega_{R+3}$.

From Lemma Ap.1 it follows that

(Ap.9) $\|\bar{D}^{M+1} S(t,\cdot;D)\|_{\Omega_{R+5},2} \leq C(1+t)^{-\beta}[\,\|u_0\|_{2,M+1} + \|u_1\|_{2,M} + |f|_{2,M,\beta}]$,

for any $t > 0$. Choose $\rho \in C^\infty(\mathbb{R}^n)$ so that $\rho(x) = 1$ if $|x| \geq R+4$ and $= 0$ if $|x| \leq R+3$. By the uniqueness theorem for Cauchy problem, we have $\rho(x)S(t,x;D) = S_0(t,x;D_0)$ where $D_0 = (0,0,-2\Sigma_{j=1}^n(\partial_j\rho)(\partial_j w)-(\Delta\rho)w)$ and $w(t,x) = S(t,x;D)$ if $x \in \Omega$ and $= 0$ if $x \in \mathbb{R}^n - \Omega$. By Duhamel's principle we have

$$\text{(Ap.10)} \quad \rho(x)S(t,x;D) = \int_0^t S_0(t-s,x;D_0'(s))ds,$$

where $D_s'(s) = (0,-2\Sigma_{j=1}^n(\partial_j\rho(x))\partial_j w(s,x)-(\Delta\rho)(x)w(s,x),0)$. Since $p(n) > 2(n+1)/(n-1)$, noting that the supports of $\partial_j\rho$ and $\Delta\rho$ are contained in Ω_{R+4} and applying Lemma Ap.2 to (Ap.10), we have by (Ap.9)

$$\|\overline{D}^{M+1}(\rho(\cdot)S(t,\cdot;D))\|_{p(n)}$$

$$\text{(Ap.11)} \quad \leqq C(M,\alpha)(1+t)^{-\alpha(1-\frac{2}{p(n)})}[\|u_0\|_{2,M+n+2} + \|u_1\|_{2,M+n+1} + |f|_{2,M+n+1,\beta}].$$

By Sobolev's inequality, (Ap.9) and (Ap.11), we have

$$\|\overline{D}^{M+1}S(t,\cdot;D)\|_{p(n)}$$

$$\leqq C(M,\alpha)(1+t)^{-\alpha(1-\frac{2}{p(n)})}[\|u_0\|_{2,M+n+2} + \|u_1\|_{2,M+n+1} + |f|_{2,M+n+1,\beta}]$$

$$\text{(Ap.12)} \quad \leqq C(M,\alpha)(1+t)^{-\alpha(1-\frac{2}{p(n)})}[\|u_0\|_{q(n),M+2n+3} + \|u_1\|_{q(n),M+2n+2} + |f|_{q(n),M+2n+2,\beta}].$$

Taking $\beta = d(n)r(n)$ and $d(n) = \alpha(1-\frac{2}{p(n)})$, we have (4.4) and (4.5) by (Ap.12), which completes the proof.

References

[1] J.G. Heywood, A uniqueness theorem for nonstationary Navier-Stokes flow past an obstacle. Ann. Scuola Norm. Sup. Pisa, 11 (1), (1979), 427-445.

[2] L. Hörmander, *Linear Partial Differential Operators*, Berlin, Springer, 1963.

[3] S. Klainerman, Global existence for nonlinear wave equation, Comm. Pure Appl. Math., 33, (1980), 43-101.

[4] R.B. Melrose, Singularities and energy decay in acoustical scattering, Duke Math. J., 46, (1979), 43-59.

[5] S. Miyatake, Mixed problem for hyperbolic equation of second order, J. Math. Kyoto Univ., 13 (3), (1973), 435-487.

[6] S. Mizohata, *Theory of Partial Differential Equations*, Cambridge UP., 1973.

[7] C.S. Morawetz, Exponential decay of solutions of the wave equation, Comm. Pure Appl. Math., 19 (4), (1966), 439-444.

[8] Y. Shibata, On the global existence of classical solutions of mixed problem for some second order non-linear hyperbolic operators with dissipative term in the interior domain, Funk. Ekva., 25, (1980), 303-345.

[9] ——— , On the global existence theorem of classical solutions of second order fully nonlinear hyperbolic equations with first order dissipation in the exterior domain, Tsukuba J. Math., 7, (1983), 1-68.

[10] ——— , On the existence and regularity theorem of Neumann problem for some hyperbolic-elliptic system, in preparation.

[11] Y. Shibata and Y. Tsutsumi, On a global existence theorem of small amplitude solutions for nonlinear wave equations in an exterior domain, to appear.

[12] ——— and ———, Global existence theorem of nonlinear wave equations in the exterior domain, Lecture Notes in Numerical and Applied Analysis, Vol. 6, 155-195, Kinokuniya/ North-Holland.

[13] Y. Tsutsumi, Local energy decay of solutions to the free Schrödinger equation in exterior domains, J. Fac. Sci. Univ. Tokyo Sec. IA, 31, (1984), 97-108

[14] ———, Doctor Thesis, Tokyo Univ., in preparation.

[15] B.R. Vainberg, On the short wave asymptotic behaviour of solutions of stationary problems and the asymptotic behaviour as $t \to \infty$ of solutions of nonstationary problems, Russian Math. Surveys, 30 (2), (1975), 1-58.

[16] W.V. Wahl, L^p-decay rates for homogeneous wave equations, Math. Z., 120, (1971), 93-106.

[17] D.V. Widder, *The Laplace Transform,* Princeton University Press, 1941.

Printed in Dunstable, United Kingdom

91338435R00134